Research Exploration: Transcendence of Research Methods and Methodology

Volume 2

EDITOR IN CHIEF

RHITURAJ SAIKIA

EDITORS

PRATISHA KUMARI

SUKHWINDER SINGH

AARON MNGUNI

LAXMINARAYANA MAROLI

ABRAHAM LINCOLN TORSU

OMKAR PRASAD BAIDYA

EDITORIAL

In the vast landscape of academic literature, there emerges a beacon guiding researchers and scholars through uncharted territories of methodology and research methods. "Research Exploration: Transcendence of Research Methods and Methodology Volume 2" stands as a testament to the evolving nature of the research paradigm and the relentless pursuit of knowledge.

This book, a sequel to its predecessor, doesn't merely build upon existing foundations; it transcends them. In a scholarly dance of innovation and intellectual prowess, the authors delve into the intricacies of research methodologies, unraveling complexities and introducing a new dimension to the research landscape.

At its core, Volume 2 is a manifesto for research transcendence. It challenges the conventional boundaries of research methods, encouraging scholars to break free from the shackles of tradition and embrace a more holistic, dynamic approach. The contributors, a constellation of seasoned researchers and fresh minds, bring diverse perspectives that enrich the discourse on methodology.

One of the striking features of this volume is its inclusivity. It accommodates a spectrum of research domains, making it a valuable resource for researchers from various disciplines. Whether you're a social scientist, natural scientist, or humanities scholar, the book offers a tapestry of methodologies, each thread woven with precision and relevance.

The chapters within this volume read like a collaborative symphony, where each contributor plays a unique instrument, contributing to the harmonious melody of advancing research methodologies. From qualitative to quantitative approaches, from case studies to experimental designs, the book spans the entire spectrum of research techniques, fostering a sense of unity in diversity.

What makes this volume truly stand out is its forward-thinking approach. It doesn't just showcase current methodologies but anticipates the future of research. By exploring emerging trends, technological integrations, and interdisciplinary intersections, the book serves as a compass for researchers navigating the uncharted waters of the 21st century.

In a world where knowledge is dynamic and ever-changing, "Research Exploration: Transcendence of Research Methods and Methodology Volume 2" is a beacon, guiding scholars toward a future where the pursuit of knowledge knows no bounds. As we embark on this intellectual journey, we are reminded that true transcendence lies not in abandoning the past but in building upon it, evolving our methodologies to illuminate the path toward greater understanding.

This volume is not just a book; it is an invitation to explore, question, and redefine the very essence of research methodology. As the pages turn, so too does the collective understanding of how we approach the quest for knowledge. It is a must-read for anyone seeking to elevate their research endeavors and contribute to the ever-expanding tapestry of human understanding.

CONTENTS

1. AUVisit: A Web-based Visitors Log and Guide Application Using QR Code for Arellano University.

Ella Mae H. Na-oha, Rhonnel S. Paculanan, Jurizz F. Cabanganan, David Christian D. Sopot, James Philip M. Rodriguez, Keith Lawrence H. Escote and Elmer Pineda

Arellano University – Pasig, **Philippines**

Abstract: Web-based technologies are becoming a crucial part of many organizations since they offer streamlined and effective solutions for their daily operations in the digital age. It has been an open resource for everyone who can access the internet with their employment, entertainment, education, government, and others. These web-based technologies use the internet to interconnect networks to their users and can store data and information in different databases like the cloud. A visitor log system is one example of a web-based technology. A visitor management system, according to studiessubstitutes the manual technique of logging or gathering a visitor's information, including personal information, the person they are visiting, and even the visitation and checkout times while visiting any facility. The researchers applied a quantitative approach to get an accurate result from the quantifiable data of the sample respondents and in developing the system the proponents used the Agile Model. The system is evaluated using ISO 25010, and according to the result of the evaluation by the respondents Male users' overall average mean is 3.6 interpreted as "Strongly Agree" while male technical respondents' overall average mean is 3.6 interpreted as "Strongly Agree". Female users' overall average mean is 3.8 interpreted as "Strongly Agree" while female technical respondents' overall average mean is 3.7 interpreted as "Strongly Agree". All respondents (users and technical) strongly agreed on the acceptability and usage of the application. As conclusion that the system can serve as a modernized approach to collecting visitor information, promote connections between the university and its visitors, improve the current logging process with the use of QR codes, and generate reports that are relevant to the institution.

Keywords: Web-based, system, visitor log system, Agile Model, ISO 25010

Introduction

Web-based technologies are becoming a crucial part of many organizations since they offer streamlined and effective solutions for their daily operations in the digital age. It has been an open resource for everyone who can access the internet with their employment, entertainment, education, government, and others. [1] These web-based technologies use the internet to interconnect networks to their users and can store data and information in different databases like the cloud.

[2] Any device such as mobile devices, desktop computers, laptops, and even tablets with an internet connection can easily access this web-based system or applications whenever or wherever possible. Developers of web-based technologies have also realized the importance of accessibility with users that have limitations, hence they have adjusted their systems that adhere to the rights of every citizen to information. [1][3] The present developments, such as websites and applications, have lessened the manual work of an individual that complies with the international standards of accessibility. [1] A visitor log system is one example of a web-based technology. A visitor management system, according to studies [4] - [5] substitutes the manual technique of logging or gathering a visitor's information, including personal information, the person they are visiting, and even the visitation and checkout times while visiting any facility. This is a reason that an institution wants to protect those who are already in their vicinity, such as employees and students, from an unauthorized person who can cause unfavourable incidents that may endanger people's lives. In this case, manual visitor information logging is time-consuming for the staff and is inefficient for tracking visitors. Natural or man-made disasters could also result in the loss or damage of hard copies of the visitor's log. Taking this into account, a visitor management system using computer technology is useful to replace the traditional approach.

Objectives of the Study: The primary reason for the study is to create a web-based visitor log management system that can change how an educational institution gathers and handles visitor information. The goals are the following: Design a web-based system to serve as a modernized approach to collecting information related to visitors of an educational institution. Generate reports from a web-based visitor log management system. Make use of QR codes to improve a web-based visitor logging procedure. Promote goodwill between an educational institution and its visitors. Use ISO 25010 standards in the evaluation of the web-based visitor log management system. Determine the profile of the respondents in terms of (a) Gender.

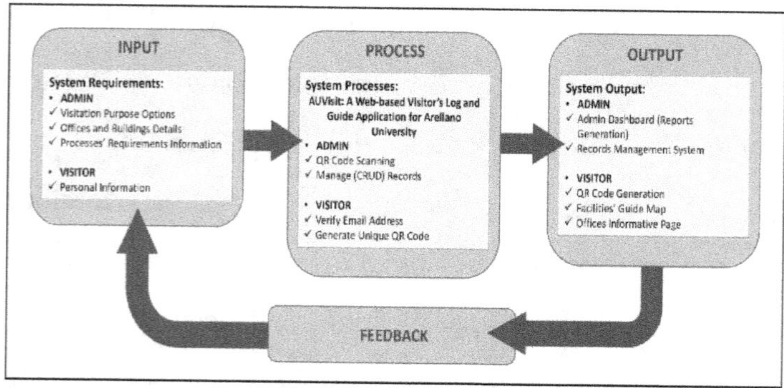

Figure 1. Input- Process- Output Model

Input: Allows admin (Security Personnel) to input the purpose of the visitation;

1. Admin (Office Staff) will provide background details of their offices and buildings;

2. Provide the requirements of each process of their respective offices;

3. The personal information of the users (visitors) is one of the requirements.

Process

1. Allows admin (Security Personnel) to scan QR Codes of the visitors;

2. Admins are allowed to manage (Create, Read, Update, Delete) the system records such as visitors, visit logs, purpose lists, office and building lists;

3. Users can generate their unique QR Codes from the personal information they provided after verifying their email addresses.

Output

1. Reports Generation can be viewed in the admin dashboards;

2. All data entered in the system are recorded and can be managed by the admin;

3. QR Codes can be downloaded by the users upon its generation;

4. A guide map for the visitors will be provided in the web application;

5. Informative pages about the university offices can be viewed by the visitors.

Significance of the Study

The web-based system project plays an important role for different types of stakeholders. It introduces a positive act of catering manual services, such as a Visitor's Log, to an internet-based platform by using technological advancements. This is very beneficial in the new normal, amidst the pandemic, where people's interactions are still limited in the country. Generally, this project helps the stakeholders in digitalizing records management as well as guiding visitors inside the institution.

The Web Application Project is seen as beneficial to the following: University Personnel: Employees, such as security personnel will also benefit from this project. Digitalizing the processes in their work will make the procedures easier and the management of visitor information more convenient. The records of the visitors are also safer inside a cloud-based database than in logbooks. **University Visitors:**The visitors in the university are usually composed of the student's parents, inquirers, and other outsiders with different purposes of visit. This web application project will be able to provide a digital identification card (ID) for these visitors, with the use of QR Codes. With this, visitor's waiting time in the institution's gateway — caused by manually writing logs – can be shortened and hassle-free. **IT/CS Professionals or Students**: These people are proficient in a variety of IT-related topics that are relevant to the subject under study in this paper. With their expertise, they will be able to understand the development process of the system, review the system structure, and give their feedback to the researchers. **Future Researchers:**This project will be important to future researchers as they study related concepts and develop similar system projects. All data and insights provided in this paper may serve as references or additional information to be able to support their research. **Research Design** The researchers applied a quantitative approach to get an accurate result from the quantifiable data of the sample respondents that will be analyzed and interpreted. **Data Collection Method** The researchers will utilize survey questionnaires and interviews as primary sources to gather necessary data. Secondary sources that are helpful in the study such as online journals, articles, and web pages are consulted and cited in the review of related literature.

System Development Life Cycle The researchers used a methodology called SDLC or System Development Life Cycle as a guide in developing the software project. It ensures the development of a high-quality system output – that is both topnotch in functionalities and aesthetics – while minimizing risks. This approach consists of seven (7) stages namely: Planning & Analysis; Define Requirements; Designing; Development; Testing; Deployment; and Maintenance [17]. **Agile SDLC Model** The Agile Model is an incremental and iterative framework that arranges SDLC phases into a more change-driven approach to software development. [17] It enables a close relationship between the developers and the stakeholders through constant feedback. The phases of the Agile SDLC Model involve Planning; Requirement Gathering; Design; Development; Testing; Implementation; and Maintenance [18]. The proponents of this study strongly believed that through this methodology, the software being developed will meet the needs of the client institution.

Figure 2. Agile SDLC Model

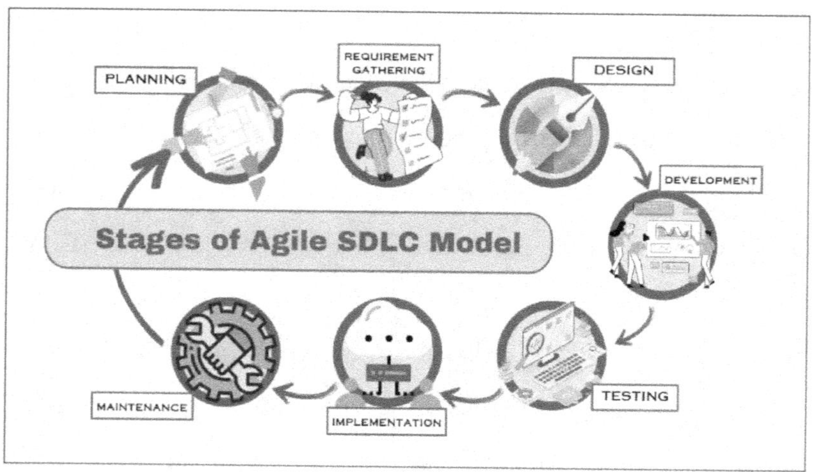

The figure shows the visualization of the different stages in the Agile SDLC framework. Starting from the planning stage to the gathering of project requirements, designing or prototyping, moving on to the developing phase, conducting testing procedures, the implementation or deployment, and towards the maintenance – before it iterates again. With this methodology, the developers will be directed throughout the whole process of software development. In this project, the planning phase involves the creation of a roadmap containing the timeline, main objectives, stakeholders, and scope of the undertaking. The researchers conducted several group meetings through various communication platforms (e.g., Google Meet, and Messenger). Requirement gathering is where the researchers need to collect information about the project, including client requirements and drawbacks, which will later define the system's features. In this stage, interviews with the people involved – security personnel – were carried out to be able to understand the visitor logs procedure. In the third stage, the software design is made about the client's needs and the essential features gathered through the preceding phase. The proponents of the system created necessary diagrams to foresee the software's result and assess the features' feasibilities. These diagrams include Wireframes for UI designs; Flowcharts for system flow; and ER (Entity-Relationship) Diagrams for the database design. The software development period involves programming with relevant languages. This is where the developers write codes for both the front-end and back-end structure of the web-based system. The next phase is called Testing which involves finding system bugs and fixing errors to ensure client satisfaction. This is also where Pilot testing, Alpha Testing, and Agile testing will be carried out for better end-product. Pilot and Alpha Testing is done to evaluate system components before their deployment, while Agile Testing is a practice that adheres to the fundamentals of Agile Software Development with its continuous cycle of development and testing [19]. The implementation stage includes the full-scale deployment of the web application project after passing all tests. The team of developers or the researchers of the study will release the software to a larger number of users (visitors and admins) who will be the source of feedback. The last phase is the maintenance of the system its functionalities, security, reliability, etc. This period covers making certain of the components updates for a smooth software experience. Maintaining the system's overall condition is very important for the software development process.

Context Diagram

Figure 3. System Context Diagram

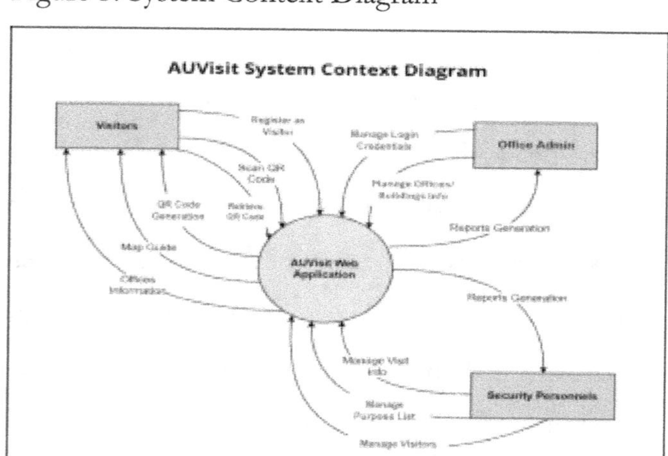

The figure above shows how the web application interacts with external entities: visitors; office admin; and security personnel. The system provides the features map guide and office information to the visitors. Also, it generates a unique QR Code for each visitor upon getting their details, which will then be scanned by the system whenever they visit the institution. The visitors are also able to retrieve their generated QR Codes in the system. The office admin of the system on the other hand manages the credentials of the admin accounts and the information about the offices and buildings that are viewed by the visitors. Lastly, security personnel manage visitors' information and the visit details, such as the visitation purpose to the institution. In addition, the system creates and presents reports to both the admin and the security personnel.

Evaluation Tools

Method for Evaluating the System

The system is evaluated using ISO 25010 which provides a standard for assessing the system's quality with the following criteria: [8]

1. Functional Suitability – The system provides functions that meet stated and implied needs when used under specific conditions (Completeness, Correctness, Appropriateness).

2. Performance Efficiency – The system's performance relative to the number of resources used under stated conditions (Time Behavior, Resource Utilization, Capacity).

3. Usability – The system can be used by specified users to achieve specified goals with effectiveness, efficiency, and satisfaction in a specified context of use (Appropriate Recognizability, Learnability, Operability, User Error Protection, User Interface Aesthetics, Accessibility).

4. Security – The system protects information and data so that persons or other products or systems have the degree of data access appropriate to their types and levels of authorization (Confidentiality, Integrity, Authenticity).

5. Portability – The system can be transferred from one hardware, software, or other operational or usage environment to another (Adaptability, Install ability, Replaceability).

Evaluation Procedure

The study's evaluation process is designed into three sections to gather necessary data from the respondents (Section 1: Socio-demographic information: Male and Female and the types of respondents, Section 2:

Technical Evaluation Questionnaire: ISO 25010 andSection 3: Additional Questions)

Respondents of the Study

The study has two groups of respondents: user and technical. These respondents are selected because of their user experiences for the user group and those with a background in information technology or exposure to system development for the technical group. The user group is made up of 70 respondents while the technical group is made up of 10 respondents for a total of 80 respondents made up of male and female genders. In selecting a sample, the study used a simple random sampling technique. The formula used to get the number of samples is called Slovin's Formula as shown below:

$$n = \frac{N}{(1 + Ne^2)}$$

Figure 7. Slovin's Formula

Respondents	No. of Respondents	Percentage
Users	60	60%
IT background/Professionals	40	40%
Total	100	100%

Table 2. Respondents of the Study

Statistical Tools

The statistical tools utilized in this study are the frequency percentage and weighted mean.

Likert Scale

The 4-Point Likert's Scale is used to analyze and interpret the data gathered from the respondents with the help of the weighted point and scale as shown in the table below:

Results and Discussion

Project Description

The output of the study is a system called AUVISIT: A Web-based Visitors Log and Guide Application Using QR Code for Arellano University – Andres Bonifacio Campus. The system uses QR codes to help the university streamline and modernize its process of collecting visitor information. The system uses a QR code that is helpful to easily track and manage visitors' information once they visit the university as it records the time-in, time-out, purpose of visitation, and the office that they want to visit as it is required before entering the premises. The system also offers to generate and download relevant reports such as daily visitor's log records, visitor's lists, and others. The system records the time-in, time-out, purpose of visit, and desired office of each visitor, and allows admins to generate and download reports such as daily visitor logs and visitor lists. The system is built using the following development tools and languages such as Hyper Text Markup Language (HTML), Cascading Stylesheets (CSS), and Bootstrap (front-end framework); JavaScript, PHP, and SQL or Structured Query Language. Agile application testing methodology is used in testing the functionality and performance of the application. Its compliance with the standards and compatibility with computer devices likewise undergo testing procedures.

Project Testing

Agile Application Testing Methodology is used to test the application. The areas covered in the testing processes are Application Functionality, Performance, Compliance, and Compatibility. The functionality is checked for errors that will affect the capabilities of the application. Performance is checked to determine the speed of loading the application on a compatible device. Compliance and Compatibility are checked if the application can run on the required hardware configuration.

Project Evaluation Results

The system is assessed in two ways: evaluations made by (a) user respondents, and (b) technical respondents. The evaluation by the user group focused on the acceptability and usability of the system as per user experiences while the other group, the technical respondents, fixated on the technical worthiness and performance of the system from their technical point of view. The two evaluation activities utilized ISO 25010 criteria.

Summary of Findings

The web application AU Visit was designed and developed to improve the management of visitors in Arellano University- Andres Bonifacio Campus. Its general objectives are to modernize the whole process of visitor management and promote the educational institution with an informative digital platform. With the integration of QR Code generation and scanning, the processes of managing visitor logs are expected to be faster and more secure. Likewise, the ISO 25010 evaluation format is used by users and technical respondents in the evaluation of the acceptability and usability of the application. This study is significant to Arellano University – Andres Bonifacio Campus, specifically for their security personnel, as the system output is expected to improve their visitor management from manual logs to a digitalized web application. In addition, the system will reduce the visitors' waiting time by manually writing logs before entering the campus facilities and guiding them throughout their visitation.

Conclusion

The study was able to identify the genders of the respondents as male and female. Evaluation results showed that all respondents, both users and technical staff regardless of their gender, strongly agreed with the acceptability, usability, technical merit, and performance efficiency of the AU Visit web application. All respondents, both male, and female, also had positive feedback, saying that the system can serve as a modernized approach to collecting visitor information, promote connections between the university and its visitors, improve the current logging process with the use of QR codes, and generate reports that are relevant to the institution.

Recommendations

The web-based system will benefit not only the university and its visitors but also future researchers through the following recommendations:

1. Integrate the system with other existing digital platforms of the institution such as their main website, learning management system, and others.

2. Enhance the project by adding a feature that collects feedback to be able to help them improve the system and overall user experience.

3. Enhance the guide map for visitors to easily locate an office that they are looking for.

4. Integrate the system as a mobile application for more convenience for the users.

5. Add more security such as two-factor authentication since visitor's information is collected by the system.

Reference

[1] Web Accessibility Initiative, "Introduction to web accessibility," w3.org/WAI/. https://www.w3.org/WAI/fundamentals/accessibility-intro/ [Accessed Jun. 19, 2023].

[2] O. Bamodu, B. Otafu, and L. Tian, *Secure web based system development*: Proceedings of the 2nd International Conference on Computer Science and Engineering (ICCSEE 2013), Atlantis Press, Paris, France, 2013

[3] S. L. Henry, S. Abou-Zahra, J. Brewer, "The role of accessibility." Core.ac.uk. https://core.ac.uk/download/pdf/78053066.pdf [Accessed Jun. 19, 2023].

[4] M. Oktaviandri and F. Kah Keat, "Design and development of visitor management system," Journal of Intelligent Manufacturing & Mechatronics, vol. 01, no. 01, 73-79, n.d. [Online]. Available:

https://core.ac.uk/download/pdf/211030682.pdf . [Accessed Jun. 20, 2023].

[5] T. Agarwal, P. Gurung, and A. Kumari, "Visitor's management system," International Research Journal of Modernization in Engineering Technology and Science, vol. 02, no. 07, July 2020. [Online]. Available: https://www.irjmets.com/uploadedfiles/paper/volume2/issue_7_july_2020/2216/1628083080.pdf . [Accessed Jun. 20, 2023].

[6] K. Feldman, "Input-Process-Output (I-P-O) Definition," iSixSigma, Nov. 07, 2018. https://www.isixsigma.com/dictionary/input-process-output-i-p-o/ . (Accessed Jun. 19, 2023).

[7] S. Schulfer, "What Is a QR Code? QR Code Meaning & Example," SproutQR, Sept. 14, 2020. https://www.sproutqr.com/blog/what-is-a-qr-code#toc-what-is-a-qr-code-qr-code-meaning . [Accessed Jun. 20, 2023].

[8] "ISO/IEC 25010," iso25000.com. https://iso25000.com/index.php/en/iso-25000-standards/iso-25010. [Accessed Jun. 20, 2023].

[9] "Visitor Management Systems: What You Need to Know," Yondu, Sep. 26, 2018. https://www.yondu.com/articles/visitor-management-systems-what-you-need-to-know/

[10] "The Key Advantages of a Visitor Management System for Schools | VisiPoint." https://www.visipoint.net/blog/key-advantages-visitor-management-system-for-schools/ . [Accessed Jun. 27, 2023].

[11] "Visitor Management QR Code," *Vpod.*

https://vpodsolutions.com/visitor-management-qr-code [Accessed Jun. 27, 2023].

[12] "QR Codes in schools, how safe are they?," Ateneo de Manila University, May 20, 2021. https://2012.ateneo.edu/udpo/article/qr-codes-in-schools-how-safe-are-they .

[13] M. Yanis, M. Zainal, R. A. Putra, and A. Y. Paembonan, "Integration of QR-Code and web-based application for developing digital tourism in Iboih Village, Indonesia as a lesson learned media on the volcanic island," GeoJournal of Tourism and Geosites, vol. 47, no. 02, 499-507, 2023. Available: https://gtg.webhost.uoradea.ro/PDF/GTG-2-2023/gtg.47217-1049.pdf . [Accessed Jun. 23, 2023]

[14] C. Jathar, S. Gurav, and K. Jamdaade, "A review on QR Code analysis," International Journal of Application or Innovation in Engineering & Management (IJAIEM), vol. 08, no. 07, July 2019. Available: https://www.ijaiem.org/Volume8Issue7/IJAIEM-2019-07-02-3.pdf . [Accessed Jun. 24, 2023].

[15] S. Arquisola, R. A. Diray, J. Dumancas, F. Habagat, B. R. Loresto, and R. Pimentel, "Visitor management information of SEAFDEC Aquaculture Department," Central Philippine University, https://repository.cpu.edu.ph/bitstream/handle/20.500.12852/2286/BSIT_ArquisolaSO_2015_Ab.pdf?sequence=1&isAllowed=y . [Accessed Jun. 23, 2023].

[16] C. Quirino, B. Sagayno, J. Cajipe, "Visitor management system in Novaliches Quezon City public library," Ascendents Asia Singapore –

Bestlink College of the Philippines Journal of Multidisciplinary Research, vol. 02, no. 01, March 2020. Available: https://ojs.aaresearchindex.com/index.php/aasgbcpjmra/article/view/2439 . [Accessed Jun. 23, 2023].

[17] H. Clark, "6 Stages Of The Software Development Life Cycle (SDLC)," The Product Manager, May 20, 2022. https://theproductmanager.com/topics/software-development-life-cycle/

[18] E. R. P. Informer, "What is Agile SDLC? (Phases, Methodologies, and Disadvantages)," www.erp-information.com, Sep. 01, 2022. https://www.erp-information.com/agile-sdlc?expand_article=1 (accessed Jun. 28, 2023).

[19] T. Hamilton, "What is Agile Testing? Process, Strategy, Test Plan, Life Cycle Example," Guru99.com, Sep. 18, 2019. https://www.guru99.com/agile-testing-a-beginner-s-guide.html

[20] PHP, "PHP: What is PHP? - Manual," Php.net, 2019. https://www.php.net/manual/en/intro-whatis.php

[21] Y. Dodge, "Weighted arithmetic mean," SpringerLink, https://link.springer.com/referenceworkentry/10.1007/978-0-387-32833-1_421 (accessed Jul. 4, 2023).

2. An Ecological analysis on William Wordsworth's "The Daffodils.

M. Julien Mary[1] and Thokchom Sunanda Devi[2]

[1]Research Scholar, Department of English, St. Joseph University, Ikishe model Village, Virgin Town, Dimapur, Nagaland, **India**

[2]Supervisor, Associate, PhD, Professor & Head of the Department of English. St, Joseph University, Dimapur. Nagaland, **India**

Abstract: William Wordsworth, as a prominent writer of the first generation English Romantic poets, wrote the remarkable poem "I Wandered Lonely as a Cloud" (1807), which is sometimes referred as "The Daffodils." This poem deeply describes the ecological concerns of the modern world, from the challenging situations that we face today to the quest of a tranquil life, by exploring the literary landscape of the Romantic period. In today's mechanically advanced world, human existence has transformed into a relentless search for work have deeply entangled in the realm of Information Technology. The simple serenity of life and the time for meaningful human connections have faded away. Our appreciation for the beauty of nature has declined due to well-equipped buildings, factories and developed cities everywhere. Today the "Daffodils" stands as a testament to the ability of the human imagination to find solace in nature, even in solitude, and the poet also uses a number of metaphors and epithets to illustrate the 'golden daffodils' which he effectively conveys his experiences, his peaceful moments, and the complete delight of his solidarity with nature. This paper will research how in the name of advancement, the natural geographies of hills, denes, and beautiful views have given way to man-made constructs, designed for human appreciation. This metamorphosis has limited openings for the admiration of nature. thus, this exploration composition, while pressing the enduring beauty and the eventuality for serenity in the ultramodern world, seeks to inspire a shift in perception by analysing Wordsworth's iconic verse," The Daffodils."

Keywords: William Wordsworth, The Daffodils, English Romantic poetry, ecological transformation, human connection, technological advancements.

Introduction

William Wordsworth, a nature poet of the Romantic period played a vital role through his poems in 18th century. His passion for nature and focus on the common man's life were enormous to the literary field that has even been studied by many researchers to teach the young minds the importance of life and nature. In one of his poems he begins with "I wandered as a lonely cloud" which is otherwise known as 'The Daffodils' worldwide. 'The Daffodils' offers alasting message of the eternal beauty of a serene life. Ecological concerns are in demand today in the contemporary world and the urban landscape and technological entanglement of contemporary life give an alarm to global threats and calls to concentrate on the pursuit of a balanced existence in the rapidly changing world in addressing issues such as climate change, human connection, and loss of nature makes us know the disastrous future. William Wordsworth (1770-1850) was a well-known English poet and ansignificant figure in the Romantic historical movement, which developed in the late 18th and early 19th centuries. He is often regarded as one of the most significant poets of the Romantic era, and his contributions to English literature were profound .He was born in the charming Lake District of England He lost his mother at a young age and was separated from his siblings due to financial difficulties in his family. These early understandings of loss and separation had a reflective impact on his poetry. During the Romantic period and Wordsworth's time, the fundamental concern towards nature and the environment was experiencing a transformation. The industrial revolution was redesigning the landscape of England, with urbanization and scientific advancements changing the relationship between humanity and the natural world. This fast industrialization led to environmental degradation, including deforestation, pollution of rivers, and the destruction of natural habitats (John,2008).

Romanticism and Revolution.

The Romantic period was an age of revolutions not only in literature but also in politics and attitude. During French Revolution that took place in 1789, not only gave great relief to numerous youthful people about the ideals of the French Revolution Equality, liberty, and

Fraternity but also caused disasters like regions of terror, thousands of bloodshed, and violence, it was also marked as a period of violence and insecurity and it had lot of political disorder and thus the 18th century was known as the age of revolutions in the History of the world indeed moment. During these circumstances, Wordsworth who was youthful at that time travelled to France to have a kind of revolutionary experience and it was the time when romantic poetry came in, he felt a lot of uneasiness after the revolutions and to escape from the situation he gave a lot of ideas to naturalness, which also marks the famous quotations of Wordsworth that poetry is *"the spontaneous overflow of powerful feelings recollected in tranquillity"* written in his 'preface' to Lyrical Ballad in 1800. Romanticism opened all entries to gain freedom of opinion and expression". The impending environmentcrisis has motivated many Romantic scholars to reconsider the Romantic's love of nature.Though it has often been mischaracterized as escapist, many writers, such as Johnathan Bates,James McKusik, Seth Reno, and others, take a point of view, arguing that Romantic poetry isactually the first instance of western proto-ecological literature. This "Green Romantic"perspective stands in stark contrast to earlier views held by new historicist scholars such asJerome McGann, Marjorie Levinson, and Alan Liu who argues that the "romantic idealizationof nature serves primarily as a mode of displacement of the political failures of the Frenchrevolution" (Carlisle2017: 1)."The romantic ecology reverences the green earth because it recognizes that we live neither physically nor psychologically can we live without green things.With the inception of the theory of the eco criticism, William Wordsworth has become theiconic figure of the theory. The connection between human beings and nature is very powerful."Romanticism poetry engages urgent issues that face us today about the relationship betweenhuman consciousness and nature, and about the structures of realization and feeling thatdispose us to act in certain ways within our environment. Rather than turn to romanticismas a guide to current environment practices, our interest is in romanticism as a site for theemergence of eco poetics and as a discourse that opens up critical questions and lines ofinvestigation about our human place in the life world" (Gary2006: 2). Indeed, Romanticism's commencement coincided with the onset of the French Revolution in 1789, challenging established systems and sociological perspectives. The revolution, characterized by the triangle ideals of liberty, equivalency, and fraternity, led to the dismantling and reconstruction of settled institutions like absolute monarchy and feudalism. This tumultuous period matched the emergence of Romanticism, marked by the publication of" Lyrical Ballad" by Wordsworth and Coleridge. Coleridge, in particular, exalted the imagination as a catalyst for exhuming the profound and new amidst the saying. This importance on life, imagination, aesthetic appreciation, and artistic prowess exemplified the Romantic period. (George, 1962)

Exploring the Poem 'Daffodils': An Overview

"I Wandered Lonely as a Cloud" (also generally known as" Daffodils") is a lyric by William Wordsworth that's known for its festivity of the beauty of nature and its capability to bring joy and happiness in times of despair. The lyric is written in the first person, with the speaker describing their own particular experience of wandering through a field of daffodils. The lyric is divided into four stanzas, each of which describes a different aspect of the speaker's experience. The first stanza sets the scene and describes the speaker's feeling of loneliness as they wander through the country. The imagery of a cloud floating over hills and denes evokes a sense of seclusion and detachment, which contrasts with the vibrant and joyous scene that the speaker is about to encounter. The second stanza describes the moment when the speaker comes across a field of daffodils. The imagery of the daffodils stretching in a" never- ending line" along the oceanfront creates a sense of admiration and wonder, and the comparison to the stars in the Milky Way emphasizes the hugeness and beauty of the scene. The use of the word" host" to describe the daffodils also suggests that they aren't just a arbitrary collection of flowers, but an systematized and purposeful group. The third stanza describes the speaker's emotional response to the scene. The use of words like" gay" and" jocund" suggest that the sight of the daffodils has brought the speaker a sense of happiness and pleasure. The expression" I gazed and gazed but little thought what wealth the show to me had brought" suggests that the speaker didn't completely realize the impact the daffodils would have on them until after they had seen them. The final stanza describes the lasting effect of the daffodils. The use of the expression" And then my heart with pleasure fills and dances with the daffodils" suggests that the sight of the daffodils has lifted the speaker's spirits and brought them a sense of joy and happiness. The lyric is also known for its use of imagery, particularly the image of the daffodils dancing in the breath, which creates a sense of movement and vibrancy that contrasts with the stillness and solitude of the opening lines.The poem also makes use of personification, as the daffodils are described as "dancing" and "tossing their heads" which imbues them with human-like characteristics and adds to the lively and joyful nature of the scene. This poem Daffodils is a very simple but a lovely and most famous poem in the Wordsworth panorama of poetry. It reminds us the familiar subjects of Wordsworth's poetry that are memory and nature. This time the poet has used a simple musicality to create eloquence in this poem. The plot of the poem is very simple. It depicts the poet's wandering and the result of this wandering emerges in the form of a beautiful cluster of dancing daffodils beside the lake. The memory of that whole picture pleases and comforts him when he is alone, gloomy and when restlessness tries to occupy him. The way the poet has characterized the occurrence of memory of the daffodils gives a strong feeling of inner satisfaction when one recalls the memory of any beloved person orany beloved object. The reverse personification of its early stanzas has the main brilliance of this poem. The speaker is compared to a natural object that is a cloud and it's the example of metaphor here as "I wandered lonely as a cloud / that floats on high..." The daffodils are continually personified as human beings, which are dancing and tossing and moving their heads in happiness. "A crowd" and "a host" are also the examples of personification.This approach creates an integral harmony between man and nature, making it one of Wordsworth's most introductory and effective styles to implant the same feeling in the anthology as the poet himself is enduring.The poet has used a good number of adjective to describe human and nature related nouns that create harmony between man and nature.

Ecological Consciousness in William Wordsworth's Poetry

For the first time the term of *ecology* in 1869 used by a great professor and philosopher *Ernst Haeckel,* it is a Greek word "oikos" etymologically means place to live, home, or household and"logy" means logical, together mean criticism. It is a study which deals with the "relations oforganisms to one another and to their physical surroundings" or the relationship between theenvironment and human beings, later on in 1878 the term of eco-criticism used in his essay"Literature and Ecology" by William Rueckert (Saman, 2017). In this world manydisasters take place on the earth, in our environment many authors think that technology and allsciences are not enough to protect nature from the ecological crises we face every day, and triedto change human behaviour towards nature by using eco criticism in literature and understand nature broadly, we cannot simply say eco criticism is just about the study of nature; it is alsorealized by ethical stands, the relationship between human beings and non-humans. There aretwo kinds of ecology: shallow ecology and deep ecology. The shallow ecology simply meansprotecting nature from pollution, to serve human beings, and remain the master of nature.However, the deep ecology means keeping nature in its original shape, without human interference of nature, or we can say the deep ecology is a movement and reflects all problemsthat will happen in the relation between human and nature. Many writers tried to have someeffect on human being's attitude towards the environment and protect nature from all pollutionscaused by industrial and commercial forces, change human beings from "ego-consciousness toeco-consciousness, and eco criticism" learn people live happy in nature by presenting theproblems and working on these problems to solve them that exist in environmental crises(Sndip, 2016). William Wordsworth was one of the poets who admired nature, description of nature was reflected in most of his poems, many critics paid attention to Wordsworth's works, especially after 1990, and they read Wordsworth's poems from his ecological estimation. Bate, Kroeber, and Mckusick are a group of critics who have positive views about William Wordsworth's ecological works and they support Wordsworth's work (Erum and Tahir,2016). Wordsworth was one of the originators of English Romantic poetry. He was considered one of the major ecological writers in nineteenth century because Romantic poets in their writings presented interconnection or the relation between humans and nature, ecologically. Subsequently, William Wordsworth's ecological point of view was reflected in his poems; he taught human beings in his writings "how to live as a part of nature" because Wordsworth and the other Romantic poets respected the green earth, especially after the industrial revolution of the eighteenth century, tried to connect human beings with nature again as that harmony has been shattered (, Zameerpal, 2017). Wordsworth's opinion about the aims of his writings or his task as a poet of nature was to use feeling and emotion to comprehend things more deeply and understanding about the "natural world" because feeling and emotion can connect man's inner self with the natural world. Wordsworth portrayed the natural world in his poems as seen in most of his works he adopted nature and remembered nature as a "source of inspiration" his experience of his life helped him to see the nature or natural world as a main power for his works, to warn human beings of all environmental problems that occurred in his time, especially after the industrial revolution. Moreover, Wordsworth believed that nature has a great power and "the force behind the world that goes through everything" (Saman2017: 830-833). It is obvious that Romantic poets were conscious of the significance of nature and may be Wordsworth comes first in his awareness of the role of nature and connect it to the life of every individual. One can see this influence through the titles of his poems and the words which are about connected to nature and individual, the two neglected ones of the previous ages. William Wordsworth, a prominent figure in the Romantic learned movement, is known for his deep ecological knowledge and connection to nature, which is beautifully reflected in his poetry. One of his most renowned lyrics," I Wandered Lonely Asa Cloud," generally known as" Daffodils," embodies his profound ecological sensibilities. Wordsworth's poetry frequently celebrates the beauty and comfort that nature offers. In"

Daffodils," he vividly describes the golden daffodils in a natural geography, emphasizing their visual appeal and the emotional impact they've on him. This festivity of nature is a vital aspect of his ecological knowledge (J.Smith, 2002). The lyric begins with the speaker feeling" lonely as a cloud," emphasizing human isolation. still, the hassle with the daffodils transforms his mood, and he feels connected to the natural world. This reflects the idea that nature provides solace and connection to the mortal spirit (Johnson, 2005). While the lyric does not explicitly argue ecological notions, it implies a collective relationship between humans and nature. The daffodils give visual pleasure to the poet, and his remembrance of the scene subsequently uplifts his spirits, illuminating the reciprocity between humans and nature(L. *Brown,* 2010).

The Contemporary Significance of 'Daffodils' in Today's World.

Today in every moment, as ecological apprehensions appear huge in our contemporary world, with urbanization and technological advancement, Wordsworth's words knock further than ever. His work serves as a call to action, encouraging us to address the continuous issues like climate change, the loss of mortal connection with nature, and the impending ecological extremity. In this analysis, we'll explore how Wordsworth's endless themes offer guidance for a balanced and sustainable actuality in our fast changing world. Abrams' work "Natural Supernaturalism Tradition and Revolution in Romantic Literature" is celebrated for its in- depth analysis of the Romantic Movement's lyrical and philosophical principles. It examines how Romantic literature responded to the intellectual and social changes of its time and how it grappled with themes related to nature, individualism, and imagination. The" Natural Supernaturalism" is a precious resource for those interested in the Romantic period, the relationship between literature and the ambient, and the disquisition of how Romantic poets like Wordsworth integrated ecological themes into their work. It provides a foundation for understanding the profound ecological connections present in the poetry of Wordsworth and his contemporaries. M.H.*Abrams* (1999). Nicholas Roe's composition" Ecocriticism Some Emerging Trends" provides a critical examination of the arising field of ecocriticism in literature. The composition discusses the evolving trends within ecocriticism, centering on the relationship between literature and the surroundings. It delves into how literary workshop, especially those from the Romantic period, have been analysed with a focus on ecological themes and perspectives. In the composition, Roe highlights the adding interest in the ecological aspects of literature, particularly in the environment of Romantic poetry. It discusses the influence of the natural world on Romantic poets like William Wordsworth and the part of nature as a source of encouragement and a central theme in their works. Roe also examines the ways in which ecocriticism has shaped the understanding of literary textbooks and their environmental applicability. The composition underscores the significance of exploring the connections between literature, nature, and ecological interests and suggests that these connections are essential for understanding the broader denunciations of literature. N.*Roe*(2004). Johnston delves into the deep and enduring connection between the Romantic poets and the natural world, concentrating on the way nature served as a profound source of comfort for their erudite works. Johnston's composition" Nature and the Romantic Poet" provides a detailed analysis of how nature is portrayed in the poetry of the Romantic period, with a particular emphasis on poets like William Wordsworth, Samuel Taylor Coleridge, and others. Johnston discusses how the Romantic poets frequently celebrated the beauty, wonder, and power of the natural world in their works. also, Johnston examines how nature served as a symbol of the poets' inner studies and passions, growing a important and emotional background for their disquisition of mortal experiences, feelings, and the sublime.K.*Johnston*(2006). ASLE is committed to advancing the study of literature from an environmental perspective, concentrating on the interplay between human culture, literature, and the natural world. The association fosters collaboration and interdisciplinary approaches, drawing from fields similar as literature, ecology, environmental wisdom, and sustainability studies.(ASLE, n.d.). After reading diverse sources about all developments took place, romanticism emerged in literature in the late of eighteenth-century and in the nineteenth century as a reaction against industrial revolution, and many writers tried to inform human beings to protect physical environment from destructions and pollution, especially William Wordsworth who was one of the pioneers who wrote about nature in particular in a simple language to convey his message to all people that nature besides its beauty and power can serve human beings psychologically and physically. Several critics described him as a great ecological poet, admirer of nature in the nineteenth century, in his most works he described nature which shows the aesthetic of Romantic poetry. However, ecological and eco-criticism found sophisticated by some groups, they organized some associations in the twentieth century to promote the relationship between human beings and the physical environment through literature to serve human beings in a safe nature. Ecocriticism has an important role in our lives because it emphasizesthe impact of nature on human beings and conversely is true. Eco-criticism can have an effective role in culture and climate by cooperating between anthropologists, writers, critics, scientists; eco-criticism can teach writers to how to work in nature in a literary work or text and ecologists encourage the writers to support human beings to have ethical treatment with nature. Eco-criticism exists in William Wordsworth's works, he wrote several poems about nature and described nature more precisely, his writings to be noticed by many critics in the twenty-first century and they are still studying his works attentively, especially "The Lucy", The Prelude", Tintern Abby", "The Tables Turned", and "My Heart Leaps Up". As the link has been demonstrated between human beings and nature, it is an inherent connection; each one can complete the other. Another important point is that human's love for nature exists in our inner that is not related to the time and age of man or the development of science and technology. Wordsworth's poems are great evidence for that. In addition, human beings could not live without nature, because human can live in urban areas for a while a long way from nature, but at the end human beings will return to nature, we are a part of nature when being alive, or dead.

Conclusion a Wake- Up Call for Ecological Awareness

As we claw into the profound ecological knowledge rooted within the poetry of William Wordsworth, particularly in the dateless verses of" I Wandered Lonely As A Cloud," it becomes apparent that the Romantic period's reverence for nature is more applicable moment than ever. Wordsworth's festivity of the beauty and alleviation drawn from the natural world serves as a poignant memorial of the significance of conserving the terrain in the face of a global trouble. In present's world, we find ourselves abruptly detached from the beauty and wonder of nature that formerly enveloped us. The daffodils, those delicate flowers untouched by scars, are fading from our geographies, echoing the broader environmental challenges we now confront.However, our unborn generations may inherit a world where exhibitions feature synthetic representations of hills, plastic clones of flowers, if we continue down this path of incuriosity. The actually coffers that nature once supplied will be replaced by unsustainable choices. It's imperative that we wake up to this impending ecological extremity, and we must act with urgency. Every moment we delay in honoring the impact of our behavior, every misstep that furthers environmental declination, blends the challenges we pass on to those who come after us. Let us make a collaborative commitment to halt the desolation of our context. Let us not stay until it's too late, for the mending of the Earth may scale generations. Our individual responsibility is our illuminate of relief. We can choose to reduce our carbon trace, advocate for sustainable practices, and educate ourselves and others about the significance of keeping our natural world. Each small action contributes to a global movement that can make a significant difference. In the spirit of Wordsworth and the Romantic poets, let us find solace and alleviation in the beauty of our surroundings. Let us appreciate the delicate daffodils, the vast geographies, and the intricate ecosystems that sustain life on Earth. Through our conduct today, we can guarantee that the beauty and vitality of the natural world endure, not only for ourselves but for generations to come.

Refernces

1. John Milton. (2008). "Industrial Revolution and Environmental Damage." Encyclopedia of Earth, edited by Cutler J. Cleveland, National Council for Science and the Environment,www.eoearth.org/view/article/153294.

2. CarlisleHuntington. (2017). "Can Poetry Save the Earth: A Study in Romantic". Washington: Puget Sound.

3. GaryHarrison, (2006). "Romanticism, Nature, Ecology". Albuquerque NM: University of New Mexico.

4. George Lefebvre. (1962.) "The French Revolution" Columbia University Press.

5. SamanMohammed, (2017). "An Eco critical Inquiry of John Keats and William Wordsworth'sSelected Poems: A Comparative Study". Vol.3, No. 2, June: 828-840.

6. Sndip Mishra. (2016). "Criticism: Study of Environmental Issues in Literature".BRICSJournalof Educational Research.Vol.6, No.4.Odisha: KIIT University.

7. Erum Sultana, and SaleemTahir. (2016). "A Manifesto of Eco-criticism". *Journal ofLiterature, Languages and Linguistics: An International Peer Reviewed Journal.*Vol.1,

8. No.19. Lahore, Punjab: University of Lahore, (7-10).

9. ZameerpalKaur. (2017). "Environmental Consciousness and Poetry: An Eco critical Studyof William Wordsworth and Bhaivir Singh's Poems". International Journal of EnglishLanguage, Literature and Translation Studies Vol.4, No.3, 410-416.

10. SamanMohammed. (2017). "An Eco critical Inquiry of John Keats and William Wordsworth'sSelected Poems: A Comparative Study". Vol.3, No. 2, June: 828-840.

11. J.Smith. (2002). "Nature and Ecology in Wordsworth's Poetry". Publisher.

12. K. Johnson.(2005). "Ecological Harmony in Wordsworth's Works." Journal of Literature and Environment, 15(2), 121-135.

13. L.Brown. (2010). "Wordsworth and the Natural World". Cambridge University Press.

14. M. H Abrams, (1999)."Natural Supernaturalism: Tradition and Revolution in Romantic Literature". W.W. Norton & Company.

15. N. Roe. (2004). Ecocriticism: Some Emerging Trends. "The Wordsworth Circle" 35(1).

16. K. Johnston. (2006). Nature and the Romantic Poet. The Cambridge Quarterly, 35(1), 1-18. doi:10.1093/camqtly/bfj001

17. Association for the Study of Literature and Environment. (n.d.). *About ASLE*. https://www.asle.org/about-asle/

3. Academic Activities Aligned with the Government Curriculum in the Preschools.

Leena Shrestha

Mphil-PhD student, Central Department of Home Science, **India**

Abstract: A preschool is a place designed for young children above 1.5 years before starting formal education or elementary school. It focuses on the holistic education of the child through various classroom hands-on activities, fun learning, and a culture of learning from early years (Teachmint, 2023). Early childhood education plays a key role in fostering cognitive and language skills, social competency, and emotional developmentand forms a baseforexcellent basic education (Unicef, 2023). Preschool age shells the two sub-stage of Early Childhood Development (ECD) years i.e.from 1.5 to 6 years. Development in this stage is the continuous procedure of obtaining skills and abilities across the different domains that helpto reason, solve troubles, express emotion, talk, and form bonds through interaction between the environment and the child. The standards set are developmentally appropriate but the academic content in the cognitive domain is a concern in most South Asian countries as it creates developmentally inappropriate loads (Unicef, 2019). In the study, the data were gathered through a scheduled questionnaire from 36 preschool teachers of 12 private preschools. The results indicated that all of the schools have government ECD curriculum and syllabus in their schools but are using syllabus (school-made curriculum according to the respondents) as curriculum. Most of the teachers are trained and find it easy to follow the book to teach the children. Grade teaching is prevalent in most of the schools and classrooms are full of copies and course books. The play materials and manipulatives are seen collected in a playroom/Montessori lab or in the office. The classroom corners are missing, and outdoor activities are limited to a few minutes of recess time. Storytelling, finger rhymes, art and craft, and field trips are some of the activities inculcated in the classes but the timing allotted for them is not followed as prescribed in the government curriculum. The written exam is used as an assessment tool. There isa lot of academic content to be covered whichresults inthe reduction of the fun-filled and recreational activities from theplanning.

Keywords: Preschools, developmental domains, academic syllabus, government curriculum, classroom activities

Introduction

Early childhood is the time from conception to eight years of age during which quick and critical growth and development take place for whichgood nurturing care is essential for children's physical, cognitive, language, and socio-emotional development(Inee, 2023).Many researches have proved that ninety percent of brain development takes place in the first five years of life. So, during this time period, it is very important that they get the right stimulation and enough opportunities to gain knowledge through exploration, play, and interaction. The connection of the neuron can be at a pace of 1000 per second which can be up to 1 million per second(Unicef, 2017). Children lacking in proper nurturing and responsive care have smaller brains and few neural connections(Inee, 2023). Young children aged 1.5 to 5.5 years get the opportunity for proper brain development in quality preschools(Teachmint, 2023). Preschools being a place for the holistic education of children should focus on the learning journey rather than the result or marks in the different subject matter. The classroom should be lively and the subject matter should be brought with the investigating, collaborating, and hands-on activities through exploration and discovery. An integrated approach to teaching should be focused rather than teaching individual subjects. Each child is unique and learnsdifferently, this should be taken positively by teachers and build the self-confidence of the learners. Help them to be persistent and resilient to form a strong base to grow and learn not only during ECD years but throughout life. Instead of memorizing the facts and getting ready for the test children in their early years should be taught cooperation and teamwork. They should be taught to think critically.(Heischools, 2022) Holistic education in preschoolssupports progress in academic accomplishments by offering a supportive learning atmosphere and catering to individual learning styles. Children build self-confidence, a sense of social responsibility, strong critical thinking skills, and other skills needed in later career life. According to the Learning Policy Institute, integrated teaching or a whole-child approach has reduced psychological impacts like violence, abuse, or poor academic achievements (SOE online, 2020).

Objectives of the Study

The general objective of the study was to identify the practices of the government curriculum in private preschools ofMadhyapurThimi Bhaktapur. The specific objective of the study is to assess the academic syllabus. To identify the academic and non-academic activities conducted in ECD classesof the private preschools of MadhyapurThimi, Bhaktapur. **Methodology** For the study,12 preschools were selected purposively where census sampling was used for the selection of 36 preschool teachers teaching in Nursery, LKG, and UKG. In-depth interview was done with scheduled interviews for collecting informationby using a structured questionnaire. Along with observation checklist was also developed for classroom inspection

Result and Discussion

The study was initiated with the belief that teachers have an idea about the governmentcurriculum and that schools follow the play-way method and work for the holistic education of children. But the study showed that the teachers do not have much idea about the government curriculum, and holistic education. The schools followed the textbook of different publications. Academically, English started from the alphabet in the Nursery and by the end of the UKG,children need to write a paragraph. Likewise, Nepali started from Devanagariletter writing (s–1) and by the end of the UKG, they need to write a paragraph. Math started with number writing, and need to write the numbers and number names up to 100. Do other mathematical operations like addition with and without carryover and subtraction with and without borrowing. The academic content was seen very high in most of the schools which was nearly equivalent to the content of grade one according to the government curriculum(CDC, 2076). Most of the schools have high content in Nursery and LKG classes. Like writing Devanagari letters i.eByanjan and sworbarna(s–1, c–cM), writing numbers up to 80/100 in Nursery. Likewise in LKG writing the words from different letters both in Nepali and English and matras-related words in Devanagari. Also, most of the teachers have anunderstanding of the syllabus as a curriculum. The obtained data were coded and calculated using descriptive statistics.

Tabe 1: Type of Curriculum used in the class room (n=36)

Type of Curriculum	Frequency	Percentage
Gov. Curriculum	6	16.67
Syllabus as Currriculum	22	61.11
Franchise	3	8.33

| Don't know | 5 | 13.89 |
| Total | 36 | 100 |

The study showed that most of the school i.e. 61.11% use the syllabus as curriculum, 16.67 % use government curriculum, 13.89% don't know which curriculum they use and 8.33% use franchise curriculum in the preschools. **Table 2: Number of subjects taught in a day (n=36)**

Number of subjects	Frequency	Percentage
1	4	11.11
2	12	33.33
3	15	41.67
4	2	5.56
5	2	5.56
6	1	2.78
Total	**36**	**100**

The study showed that 41.67% teach 3 subjects, 33.33% teach two subjects, 11.11% teach one subject, 5.56% teach four and five subjects and 2.78% teach six subjects in a day. The study showed that 58.33% of preschools use written exams as assessments. 33.33% use Portfolio and 8.33% use Continuous Assessment Systems (CAS) in the classrooms.

Conclusion: Only a few preschools follow the activity-basedplay-way method where children get a chance to do hands-on activities whereas most of the schools have a traditional method of teaching. Manipulatives, other play materials, teaching aids, and the corner setup are rarely seen in the class. Most of the classes are full of desks and benches. Activity-based teaching, exploration activities, and collaboration are rarely seen. Children are doing lots of writing in all the subjects like Nepali, English, Math, and Theme. The academic syllabus of all the classes in preschool is high which creates developmentally inappropriate loads for children. Writing before the age of 3 years when their fine motor skills are not developed properly.Most of the school hours of children are spent in the classroom. The time period divided between the school or the class routine prescribed by the government is not followed(CDC, 2076). They have their own routine in which the academic activities are mainly focused and other play-based and ECA activities are neglected. The syllabus of UKG in most schools is nearly similar to the grade one syllabus. Most of the teachers said that they have lots of content to cover and that they are forced to reduce the fun-filled and recreational activities from their planning. The achievements of the children are measured by the written exams in most of the preschools. The focus is on cognitive development and other domains of development are neglected. Even the height and weight of the children are not measured in most preschools. This indicates that there is no balance learning between all the learning areas that lead to holistic education. Physical and mental development can be measured and tested but to promote social and emotional development one should get an opportunity to explore and express their ideas and feelings which is lacking. The expectation of the parents from the school is also very high regarding academics. Parents seek high academic achievements from their ward/s. They expect more homework and ask the teachers to send more writing work so that their children will be busy at home as well and not do any kind of mischief. Most of the parents are not that well educated which also affect the expectation of parents from the school. The concept of holistic education is in the brochures and speech only. Parent awareness is minimal in this area. Even the school owners are not so aware of teacher training and new methodologies.The parents are even not aware of collaborating with schools and teachers for the development of their own children. Parent awareness programs are highly in need in this area.

Recommendations

1. The teacher and child ratio should be well managed.
2. ECD classes should be well-equipped and an ECD curriculum should be mandatory.
3. ECD classrooms should focus on 21st-century learning skills.

4. The facilitator should be at least three months ECD trained with child psychology classes.

5. The textbooks should be removed, and monitoring and supervision from the government side should be done.

6. School leaders should educate parents/ guardians about the right way of guiding children at home.

7. Parent awareness programs should be introduced.

8. School timing should not be more than four or five hours.

9. There should be age-appropriate academic activities, and the use of real objects or materials to teach. Written tests for children should be avoided.

10. Children should be encouraged to think out of the box instead of rote learning.

References

1. *Unicef.* (2019, November). Retrieved from www.unicef.org: https://www.unicef.org/rosa/reports/mapping-early-childhood-development-standards-and-good-practices

2. *Unicef.* (2018). Retrieved from www.unicef.org: https://www.unicef.org/sites/default/files/2019-05/Early%20Childhood%20Development%20in%20the%20UNICEF%20Strategic%20Plan%202018

3. CDC. (2076). *MoECDC*. Retrieved from www.moecdc.gov.np: https://moecdc.gov.np/en/curriculum

4. *Inee*. (2023, August 2). Retrieved from www.inee.org: https://inee.org/collections/early-childhood-development

5. *Heischools*. (2022, March 31). Retrieved from www.heischools.com: https://www.heischools.com/blog/holistic-education-versus-traditional-education-during-the-early-years#:~:text=The%20biggest%20difference%20between%20the,taught%20as%20one%20interrelated%20whole.

6. *Unicef*. (2017, September). Retrieved from www.unicef.org: https://www.unicef.org/media/48886/file/UNICEF_Early_Moments_Matter_for_Every_Child-ENG.pdf

7. *Teachmint*. (2023, July 21). Retrieved from www.teachmint.com: https://www.teachmint.com/glossary/p/preschool/

8. *Unicef*. (2023, July 21). Retrieved from www.data.unicef.org: https://data.unicef.org/topic/early-childhood-development/early-childhood-education/

9. *SOE online*. (2020, May 13). Retrieved from www.soeonline.american.edu: https://soeonline.american.edu/blog/what-is-holistic-education/

4. Ethical Hacking: Balancing Security and Responsibility in the Digital Age.

Shruti Garg

Assistant Professor, Department of Computer Applications, **India**

Abstract: In an increasingly interconnected and digitized world, the significance of cybersecurity cannot be overstated. The rise of cyber threats, data breaches, and vulnerabilities has necessitated the emergence of a unique breed of professionals - ethical hackers. This research paper delves into the realm of ethical hacking, shedding light on its pivotal role in safeguarding digital ecosystems while navigating the complex landscape of ethical responsibility [1]. This research paper meticulously delineates the ethical hacker's code of conduct, emphasizing the moral compass that guides their actions. It unravels the intricate tapestry of ethical dilemmas faced by ethical hackers in their relentless pursuit of security. By examining the benefits and limitations of ethical hacking, it underscores the need for a judicious balance between security imperatives and ethical principles. Drawing from real-world case studies, this paper showcases how ethical hackers have thwarted cyber threats, preserved data integrity, and safeguarded critical infrastructure. It underscores the indispensable role of ethical hacking in the organizational context, elucidating how businesses and institutions can harness its potential to fortify their digital arsenals. In an era defined by digital vulnerabilities and incessant threats, ethical hacking emerges not merely as a practice but as a moral imperative. This paper navigates the intricate terrain of ethical hacking, illuminating the path toward a safer and more secure digital future.

Keywords: Cybersecurity, Ethics, Law, Trust, Values, Hacking, Hacker

Introduction

Define Ethical Hacking: Ethical hacking, also known as penetration testing or white-hat hacking, refers to the authorized and deliberate simulation of cyberattacks on computer systems, networks, or applications to identify and rectify potential security vulnerabilities [2]. **Explore Ethical Hacking Methodologies:**Ethical hacking methodologies, also known as penetration testing methodologies, provide a systematic and structured approach to identifying and assessing security vulnerabilities in computer systems, networks, applications, and other digital assets. Ethical hackers, or penetration testers, follow these methodologies to ensure comprehensive testing and analysis. **Establish Ethical Guidelines:**Key ethical guidelines for ethical hacking are Authorization and consent, Legal compliance, Transparency and Communication, Respect for privacy, Avoiding Damage, Documentation, Confidentiality, Responsible Disclosure, Professionalism, Continuous Education and Skill Development and No exploitation for Personal Gain. **Highlight Benefits and Limitations:**

Discuss the advantages and disadvantages of ethical hacking, emphasizing its significance in improving cybersecurity while acknowledging potential ethical dilemmas and challenges.

Provide Real-world Examples: Offer case studies and real-world examples of ethical hacking incidents to illustrate how ethical hackers have contributed to resolving security issues and safeguarding digital assets. **Examine Organizational Integration:** Explore the role of ethical hacking in businesses and organizations, including how they can establish and maintain ethical hacking programs to enhance their cybersecurity posture [3]. **Predict Future Trends:** Predict future trends and challenges in ethical hacking and cybersecurity, helping readers understand the evolving landscape of digital security. **Emphasize Ethical Responsibility:** Highlight the importance of ethical responsibility in hacking practices, emphasizing the need for a responsible and ethical approach to security testing and vulnerability disclosure. **Contribute to Knowledge:** Contribute to the existing body of knowledge in the field of ethical hacking and cybersecurity by providing insights, analysis, and recommendations for ethical hacking practices. **Promote Ethical Values:** Promote and advocate for ethical values, transparency, and accountability in the context of hacking and cybersecurity. **Introduction to Ethical Hacking and Cybersecurity: Define Ethical Hacking and its importance in cybersecurity:** Ethical hacking, also known as white-hat hacking or penetration testing, is the practice of intentionally probing computer systems, networks, and software applications to identify vulnerabilities and weaknesses in a lawful and responsible manner. Ethical hackers, often referred to as penetration testers or security researchers, use the same techniques and tools as malicious hackers, but their activities are authorized and performed with the explicit goal of improving cybersecurity [4]. **Importance of Ethical Hacking In Cybersecurity: Identifying vulnerabilities:** Ethical hackers play a crucial role in discovering security flaws and vulnerabilities in a system or network before malicious hackers can exploit them. This allows organizations to proactively address these issues and enhance their security measures. **Prevention of data breaches:** By finding and addressing vulnerabilities, ethical hacking helps prevent data breaches, which can be costly and damaging to an organization's reputation. It can save companies from potential financial losses and legal liabilities [5]. **Compliance and regulation adherence:** Many industries and organizations are subject to various cybersecurity regulations and compliance requirements. Ethical hacking can help ensure that these standards are met and maintained, helping organizations avoid legal penalties and fines. **Continuous improvement:** Ethical hacking is an ongoing process. Regular security assessments and penetration testing help organizations continuously improve their security posture, adapt to evolving threats, and stay ahead of potential attackers. **Security awareness:** Ethical hacking activities raise awareness among employees and stakeholders about cybersecurity threats and best practices. This, in turn, helps create a security-conscious culture within an organization

[6]. **Risk management:** Ethical hacking provides valuable insights into an organization's risk profile. It allows decision-makers to prioritize security investments and allocate resources where they are most needed. **Incident response planning:** Ethical hackers can assist in the development of incident response plans by identifying potential attack vectors and helping organizations prepare for potential security incidents. **Highlight the growing significance of cybersecurity in the digital age:** The growing significance of cybersecurity in the digital age is driven by a multitude of factors, as our reliance on digital technologies continues to expand. Here are some key reasons why cybersecurity is increasingly crucial: **Pervasive digitalization:** In the digital age, virtually every aspect of our lives is influenced by technology. From personal communications to critical infrastructure, businesses, healthcare, and government operations, the vast majority of activities and systems are now digital, making them vulnerable to cyber threats. **Data proliferation:** The volume of sensitive and valuable data being generated, processed, and stored has surged. Protecting this data from theft, manipulation, or destruction is paramount, as data breaches can lead to severe financial and reputational damage [7]. **Evolving threat landscape:** Cyber threats are constantly evolving, with hackers becoming more sophisticated, organized, and persistent. The rise of state-sponsored attacks, ransomware, and zero-day vulnerabilities has heightened the need for robust cybersecurity measures. **Internet of Things (IoT):** The proliferation of IoT devices has expanded the attack surface. These devices, often with limited security controls, can

be exploited to gain access to networks, further emphasizing the importance of securing interconnected systems.

Future Challenges

Skill Shortages: The demand for skilled ethical hackers is likely to continue growing, leading to challenges in recruiting and retaining qualified professionals. Addressing the skills gap will be crucial for effective cybersecurity. **Regulatory Changes:** As new technologies emerge, regulatory frameworks may evolve. Ethical hackers will need to stay informed about changing legal requirements and compliance standards. **Privacy Concerns:** Increasing awareness and regulations surrounding privacy may impact ethical hacking practices. Striking a balance between security assessments and privacy considerations will be an ongoing challenge. **Global Collaboration:** Cyber threats are global, and collaboration between ethical hackers, organizations, and governments worldwide will be essential. Overcoming legal and geopolitical barriers to information sharing is a challenge. **Evolving Attack Vectors:** Malicious vectors constantly innovate their attack techniques. Ethical hackers must anticipate and adapt to emerging threats, including those exploiting social engineering and human factors. **Ethical Dilemmas:** Ethical hacking often involves navigating complex ethical dilemmas. As technologies advance, professionals may face new dilemmas related to disclosure, responsible reporting, and the potential consequences of their actions. **Integration with DevSecOps:** The integration of security into the DevOps process (DevSecOps) is a growing trend. Ethical hacking methodologies will need to align with agile development practices and continuous integration/continuous deployment (CI/CD) pipelines.

Conclusion

The field of ethical hacking stands at the forefront of safeguarding digital landscapes in an era marked by relentless cyber threats and rapid technological advancements. Through the systematic application of penetration testing methodologies, adherence to ethical guidelines and codes of conduct, and compliance with legal and regulatory frameworks, ethical hackers play a pivotal role in fortifying the resilience of organizations against malicious intrusions. As we consider the future trends and challenges in this domain, it becomes evident that the landscape is dynamic and continually evolving. The integration of artificial intelligence, the emergence of quantum computing, and the proliferation of IoT devices pose new challenges that ethical hackers must confront. At the same time, the persisting issues of skill shortages, privacy concerns, and global collaboration underscore the importance of a holistic and adaptive approach to cybersecurity. Looking forward, ethical hacking will undoubtedly remain a cornerstone of cybersecurity strategies. The ongoing collaboration between ethical hackers, industry stakeholders, and regulatory bodies will be essential in shaping a secure digital future. As technologies advance, ethical hacking practices must evolve in tandem, embracing innovation, ethical considerations, and a commitment to continuous improvement.

References

1. Anderson, R. (2001). Why information security is hard—an economic perspective. In Proceedings of the 17th Annual Computer Security Applications Conference (ACSAC) (pp. 358-365).

2. Arun, P., Shiva, S., & Mathew, L. (2016). Heartbleed bug: Vulnerability analysis and exploitation. Procedia Technology, 25, 56-63.

3. Bort, J. (2017). Equifax's data breach costs the CEO his job. Business Insider. Retrieved fromhttps://www.businessinsider.com/equifax-breach-lead-to-ceo-richard-smith-retirement-2017-9

4. Clarke, R. A., & Knake, R. K. (2010). Cyber war: The next threat to national security and what to do about it. HarperCollins.

5. EC-Council. (2021). Certified Ethical Hacker (CEH). EC-Council. Retrieved from https://www.eccouncil.org/programs/certified-ethical-hacker-ceh/

6. Facebook. (n.d.). Facebook Bug Bounty. Retrieved from https://www.facebook.com/whitehat

7. Goodman, S. E. (2015). The Ethics of Stuxnet. Philosophy & Technology, 28(1), 45-56.

8. Kim, D. S., Heo, G., Jang, Y. J., & Lee, J. (2015). An analysis of social engineering and how to prevent it. Procedia Computer Science, 60, 253-260.

9. Krebs, B. (2011). Sony breach may have exposed customer card data. Krebs on Security. Retrieved from https://krebsonsecurity.com/2011/04/sony-breach-may-have-exposed-customer-card-data/

10. NIST. (2017). Framework for improving critical infrastructure cybersecurity. National Institute of Standards and Technology (NIST). Retrieved from https://nvlpubs.nist.gov/nistpubs/CSWP/NIST.CSWP.04162018.pdf

11. Osterweil, E. (2014). Hacking Back: When Is It a Good Idea to Strike Back Against an Attacker? IEEE Security & Privacy, 12(4), 69-74.

12. Sanger, D. E. (2012). Obama order sped up wave of cyberattacks against Iran. The New York Times. Retrieved from https://www.nytimes.com/2012/06/01/world/middleast/obama-ordered-wave-of-cyberattacks-against-iran.html

13. Schneier, B. (2012). Liars and outliers: Enabling the trust that society needs to thrive. Wiley.

14. Schneier, B. (2015). Data and Goliath: The hidden battles to collect your data and control your world. W. W. Norton & Company.

15. Shamoo, A. E., & Resnik, D. B. (2015). Responsible conduct of research. Oxford University Press.

16. Singh, N., & Chaurasia, S. (2018). An analysis of Panama Papers: Cyber security and responsible hacking. Procedia Computer Science, 132, 704-709.

17. Sullivan, B. (2017). Equifax breach: 6 major missteps. Bank Info Security. Retrieved from https://www.bankinfosecurity.com/equifax-breach-6-major-missteps-a-10291

18. The White House. (2013). Presidential Policy Directive -- Critical Infrastructure Security and Resilience. Retrieved from https://obamawhitehouse.archives.gov/the-press-office/2013/02/12/presidential-policy-directive-critical-infrastructure-security-and-resil

19. Zetter, K. (2011). An unprecedented look at Stuxnet, the world's first digital weapon. Wired. Retrieved from https://www.wired.com/2014/11/countdown-to-zero-day-stuxnet/

20. Zmijewski, E. (2014). Heartbleed: A look at the numbers. APNIC Blog. Retrieved from https://blog.apnic.net/2014/04/14/heartbleed-a-look-at-the-numbers/

5. Impact of Demographic Factors on the Adoption of Internet Banking Services in the Indian Context.

Neha Dixit

Assistant Professor, Department of Business Administration, Maharana Pratap Engineering College, Kanpur, **India**

Abstarct: In this era of innovative technology advancement, every banks are trying to take advantage of the advanced technology and wants to adopt it to improve its services for its customers. The purpose of this research was to examine the impact of demographic factors on adoption of Internet banking service in India. An empirical study has been carried out with the help of 700 respondents who are using banking services. The primary data was collected through a well-structured questionnaire. Descriptive statistics was used to explain the demographic profile of the respondents. Correlation and Regression analysis is used to find out positive or negative relation between demographic factors and adoption of internet banking services. The finding depicts that age, gender, education and income has significant impact on usage of internet banking services. The significance of this research is to provide information to the banks so that banks could improve communications strategies to increase the adoption rate of internet banking services.

Keywords: Internet banking, Demographic, Customer adoption

Introduction

In the rapidly evolving landscape of financial services, the widespread adoption of internet banking stands as a pivotal phenomenon. Banks are always in favour of adoption of innovative technology. Nowadays by use of internet banking banks are increasing their customers, operational efficiency and improve their service quality. Gradually, customers are also realising that internet banking is very useful and time saving mode.Without the availability of internet access, customers will not be able to make use of the internet banking services offered by banks. Mols (2000), Internet banking makes it possible for banks to offer consumers a variety of services 24 hours a day. An Internet bank may offer general and customer-specific information, the ability to conduct transactions, access to a variety of interactive financial calculators and worksheets, customization of the content of the Internet bank and the messages sent from the bank, and finally it is possible to interact with a bank adviser, usually via e-mail, but video based advisory services via the Internet are also technically possible.

This research aim to explore the influence of demographic factors on individuals utilization of internet banking services. As digital platforms redefine the ways to manage finances, understanding how age, income, education and other demographics shapes the adoption of online banking become paramount. This investigation seeks to unravel patterns, preferences and barriers, shedding light on the intricate interplay between demographics and internet banking usage. Though this exploration, we strive to contribute valuable insights to contribute valuable insights to financial institution and policymaker navigating the dynamic terrain of digital banking.

Need of the Study

The most recent delivery channel to be introduced is internet banking in the world of banking. Internet banking is useful for banks as well for customers. There are lot of drastic changed has been seen in the banking services. In the past, a number of researches have been conducted to determine the factors affecting adoption of internet-banking. But the need to investigate the impact of demographic factors on the adoption of internet banking services is underscored by dynamic nature of consumer behavior in the digital era.

Understanding how age, income, education and other demographic variables influence the resistance to online banking is crucial for baking institutions and policy makers. This research seeks to address this need by providing insights that can inform targeted strategies, user-centric design. The aims to bridge the gap between technological advancement and the diverse preferences and constraints of different demographic groups, fostering a more inclusive and accessible landscape for internet banking services.

Objectives of the Study

To study the consumer bahavior towards adoption of internet banking services in India following objectives have been framed:

1. To explore and analyse the demographic profiles of internet banking users and non-users.
2. To identify the patterns related to age, Income, education and other relevant factors.

Review of Literature and Theoretical Framework of Hypotheses

Aysha Fathima, (2022) investigate the effect of user's demographics on factorsinfluencing their acceptance towards this banking technology.Results depict a positiveeffect of demographics on acceptance. Male users perceive them to be beneficial thanfemale users. Married individuals are found to perceive more value while using internetbanking. Young adults are using internet banking more efficiently rather than old. Higher Income group people has the most significant influence on determinants of internet bankingadoption. these group shows the increased level of internet banking adoption.

Chauhan et.al., (2016) studied on Demographic Influences on technology adoption behavior and found that except marital status many factors are which directly or indirectly influence customer behaviour towards adoption of internet banking. Out od all factors age is one of the most importance factors which has significantly influence on customer bahavior. This study also identified min factors oftechnology acceptance model (TAM) such as perceived usefulness, perceived ease of use, intention to use, and attitude and included additional factors such as social norms and perceived risk to measure e-banking adoption behavior. Margaret and Ngoma, (2013) researched on socio demographic factors of internet banking and found that age, occupation, income, gender and educational level had positive relationships with internet banking adoption. The young generation tend to be the early adopters whilst the aged tend to be the laggards hence marketing to the aged should be aimed at creating awareness. For the affluent that are quick to adapt to new technology marketing campaigns should be centred on maintaining interest. Dhanya and Velmurugan, (2021), examine the important influence of demographic factors on customer fulfilment. Research revealed that higher-educated, high income clients who utilize Internet banking for much of their financial accounts and who have been using it for extended periods appear to has more positive views and attitudes against e-banking services. In this regard, banks requirement to take into account their consumers' educational level, their income level, their general internet association, and also their expertise with e-banking services. Ameme, (2015), investigate Impact of Customer Demographic Variables on the Adoption and Use of Internet Banking in Developing Economies. This study has provided perspectives on internet banking in relation to factors that contribute to customers' adoption and usage of internet banking services. The restriction of the study to a commercial bank from Ghana may limit generalisation. The findings, however, contribute to the understanding of the demographic factors affecting the usage and adoption of internet banking services.

Onyia and Tagg (2011), Investigated the effect of demographic factors on banks customers attitudes and intension towards banking adoption in a major developing African country. The researcher have found that there are variables were correlated with attitude and intention, only gender, level of education, and employment status showed significant ability to influence Nigerian customers' attitude and intention toward IB adoption. Sohail and Shanmugham (2004) wrote a research paper concerningcustomers' preference towards e-banking in Malaysia. They concluded that age andeducational qualifications of electronic and conventional banking have no significantimpact on e-banking adoption, they further argued that accessibility to the Internet,awareness of e-banking and customerresistance to change are the main factorsaffecting adoption of e-banking. Hill (2004) conducted a study concerned with identifying the characteristics ofonline banking users. She states that it is commonly assumed that demography doinfluence the acceptance of electronic self-service tools, such as online banking. Theresults of the study have been exhibit that people who use such services are young, techno savvy and high earning. They frequently use online banking tools and theywant to conduct all transactions through online banking. There are different factors which have emerged from literature reviewed, following hypothesis have been developed in the present study and are tested in an empirical manner in order to see if these factors are effectively influencing i-banking adoption or not.

H1: Gender has a significant influence on adoption of internet banking services.

H2: Age has a significant influence on adoption of internet banking services.

H3: Education has a significant influence on adoption of internet banking services.

H4: Occupation has a significant influence on adoption of internet banking services.

H5: Income has a significant influence on adoption of internet banking services.

Research Methodology

In this empirical research study, quantitative and qualitative techniques have been used. In the Qualitative techniques secondary method has been used to collect information. This informationis based on review of literature, newspaper, articles etc. In the quantitative techniques primary method has been used to collect information about bank customers. The main instrument of primary data collection is questionnaire which have certain questions. Questionnaire has two sections; Section A describe basic characteristics of respondents and Section B explain demographic factors which influence customer behavior towards internet banking adoption. Random sampling was used to collect responses form the respondents. The questionnaire were distributed to 700 respondents but only 450 responses were collected truely. The data were collected from metropolitan cities in India.Statistical tools of SPSS is used for data inputs and analysis. Then the statistics results have been presents and analyzed.

Result Analysis and Discussion

A cross-tabulation describes two or more variables simultaneously and results in tables that reflect the joint distribution of two or more variables that have a limited number of categories or distinct values. The categories of one variable are cross-classified with the categories of one or more other variables. In our research we are interested to determine whether Internet usage is related to gender, age, income, occupation, education and level of computer literacy. The cross-tabulation is shown in Table 1. A cross tabulation includes a cell for every combination of the categories of the two variables. The number in each cell shows how many respondents gave that combination of responses. Gender may be considered as the independent variable and internet users as a dependent variable and calculating percentage is as shown in Table 1. Note that whereas 56.7 percent of the males are Internet users, only 28.2 percent of females fall into this category. This seems to indicate that males are more likely to be users of internet banking as compared to females. The calculated chi-square statistic had a value of 30.191, which is significant at the 0.05 level.So the result suggested that H1 hypothesis is accepted.

Table 2 Users * Age Cross-tabulation

			Age					
			18-25	26-35	36-45	46-55	56-65	Total
Users	Yes	Count	33	119	37	23	6	218

			23.2%	54.3%	86.0%	57.5%	100.0%	48.4%
		% within Age	23.2%	54.3%	86.0%	57.5%	100.0%	48.4%
	No	Count	109	100	6	17	0	232
		% within Age	76.8%	45.7%	14.0%	42.5%	.0%	51.6%
Total		Count	142	219	43	40	6	450
		% within Age	100.0%	100.0%	100.0%	100.0%	100.0%	100.0%

Chi-square=71.207, Sig. p < 0.05

In the Table 2, age may be considered as the independent variable and users of internet banking as a dependent variable. Mature customers or adult customers having more percentage on internet banking users which is 54.3 percent and 86 percent respectively 26-35 and 36-45 age group customers. While as same more number of respondents (76.8 percent) which have age category 18-25 is not using internet banking. This seems to indicate that mature customers are more likely to adopt internet banking as compare to young customers. The calculated chi-square (x^2) had a value of 71.207 which is significant (p) at 0.05 levels.So result analysis suggested that H2 hypothesis is accepted.

Table 3 Users * Education Cross-tabulation

			Education				Total
			Intermediate	Bachelor's	Master's	Professional	
User	Ye	Count	2	15	122	79	218

s	s	% within Education	14.3%	18.8%	49.4%	72.5%	48.4%
	N	Count	12	65	125	30	232
	o	% within Education	85.7%	81.3%	50.6%	27.5%	51.6%
Total		Count	14	80	247	109	450
		% within Education	100.0%	100.0%	100.0%	100.0%	100.0%

Chi-square=60.079, Sig. p < 0.05

Education may be considered as the independent variable and internet users as a dependent variable and calculating percentage is as shown in Table 3. Note that whereas 72.5 percent of the professional fall in the users of internet banking, as compared to Intermediate (14.3 percent), bachelor's (18.8%) and master's (49.4 percent). On the other hand, 85 percent of the intermediate and 81.3 percent of the bachelor's respondents shows attitude as a non-users. In the case those customers who are having master's degree their percentage are much closer for users and non-users, for the users 49.4 percent and for non-users 50.6 percent respectively. Hence more educated customers are more prefer to use internet banking which is significant (p) at 0.05 level and had chi-square (x^2) value 60.079. Hence H3 hypothesis is supported.

Table 4 Users * Occupation Cross-tabulation							
			Occupation				
			Private Service	Government service	Professional	Business man	Others	Total
Users	Yes	Count	58	25	70	24	41	218
		% within Occupation	32.4%	46.3%	63.6%	66.7%	57.7%	48.4%
	No	Count	121	29	40	12	30	232
		% within Occupation	67.6%	53.7%	36.4%	33.3%	42.3%	51.6%
Total		Count	179	54	110	36	71	450
		% within Occupation	100.0%	100.0%	100.0%	100.0%	100.0%	100.0%

Chi-square=35.955, Sig. p < 0.05

Occupation may be considered as the independent variable and internet users as a dependent variable and calculating percentage is as shown in Table 4. The table shows that 66.7 percent of those who are businessman and 63.6 percent those who are professional and 57.7

percent others are using internet banking, as compared to private service (32.4 percent), government service (46.3 percent). In the same case also private service, government services are not using internet banking. We were tempted to conclude that Occupation has significant impact on customer attitude towards internet banking uses. The calculated chi-square (x^2) had a value of 35.955 which is significant (p) at 0.05 level. Hence H4 hypothesis is accepted

Table 5 Users * Income Cross-tabulation								
			Income					
			< 15000	15001-35000	35001-55000	55001-75000	> 75000	Total
User s	Ye s	Count	21	52	54	62	29	218
		% within Income	22.1%	36.4%	55.7%	72.9%	96.7%	48.4%
	N o	Count	74	91	43	23	1	232
		% within Income	77.9%	63.6%	44.3%	27.1%	3.3%	51.6%
Total		Count	95	143	97	85	30	450

Table 5 Users * Income Cross-tabulation								
			Income					
			< 15000	15001-35000	35001-55000	55001-75000	> 75000	Total
Users	Yes	Count	21	52	54	62	29	218
		% within Income	22.1%	36.4%	55.7%	72.9%	96.7%	48.4%
	No	Count	74	91	43	23	1	232
		% within Income	77.9%	63.6%	44.3%	27.1%	3.3%	51.6%
Total		Count	95	143	97	85	30	450
		% within Income	100.0%	100.0%	100.0%	100.0%	100.0%	100.0%

Chi-square=85.126, Sig. p < 0.05

The Table 5 shows that 96.7 percent respondents who have monthly income more than 75000 and similarly 55.7 percent respondents who have monthly income 55001-75000 they prefer to use internet banking after that those respondents who have monthly income 35001-55000 (55.7 percent) also prefer to go for internet banking, as compared to lower income. Those respondents who have monthly income less than 15000 and 15001-35000 they never use internet banking. As a result, we found that higher income group has a significant impact on attitude towards users of internet banking which had chi-square (x^2) value of 85.126 and significant (p) at 0.05 level. The result concluded that H5 hypothesis is accepted.

Conclusion and Suggestions

From the study it can be concluded that Internet Banking is becoming the essential part of modern-day banking services and the banks are making every effort to make internet banking a success. In conclusion, the adoption of internet banking is intricately woven into the fabric of demographic factors, with gender, age, education, occupation and income emerging as pivotal influencers. Our comprehensive analysis highlights the nuanced ways in which these variables shape individuals' choices and interactions with digital financial services. Recognizing the significance of gender-based preferences, the evolving impact of age on technological adaptability, the role of education in fostering digital literacy, occupational demand influencing accessibility, and income level serving as determinants of financial inclusion, It has been observed in the present study that female respondents are less inclined to use internet banking. So, efforts must be made to encourage male bank customers to use this service. Many customers who are mature and adult are using internet banking but those have young generation they are less inclined to use it. Banks should adopt strategies so that young generation can shift to intent banking without trust factor.Banks should promote internet banking features and demonstrate how to use net banking by conducting seminar, workshop, etc and educate more and more customers so it will easy to adopt internet banking.

Limitations and Scope of Future Research

Many limitations were faced while conducting this study. The data were collected only from metropolitan cities like Delhi and Lucknow. So based on this finding we cannot judge the consumer behavior towards internet banking adoption for whole country. Second limitation is that the responses were only 450 which is less as compare to population of India.A scope of further research that research can investigate how cultural factors influence internet banking adoption, going beyond to demographic factors to explore the impact of cultural values and norms. Nowadays technological advancement and innovations has been seen. So the researcher can examine the impact of emerging technologies like AI on adoption of banking services and how different demographic factors respond to these innovations.

Implication of the Study

The study will be a benefit to the banks. The information will be helpful to researchers, bankers, policy makers and the Government, as the development of internet banking. It will great help for development and growth of the country.

References

1. Aysha Fathima (2022), Influence of Demographic Variables on Determinants of Internet Banking Adoption, ECS Transactions, Article

2. Chauhan, V., Choudhary, V., and Mathur, S., (2016), Demographic Influences on Technology Adoption Behavior:A Study of E-Banking Services in India, Prabandhan Indian Journal of Management, Vol. 9, Issue 5, pp. 45.

3. Mols P., N., (2000), The Internet and service Marketing – The case of Danish retail banking, Internet Research, Vol 10, Issue 1, pp. 7-18.

4. Margaret, M. and Ngoma, N. F., (2013), Socio-Demographic Factors Influencing Adoption of Internet Banking In Zimbabwe, Journal of Sustainable Development in Africa, Vol. 15, Issue. 8.

5. Dhanya, B. K., and Velmurugan, V. P., (2021), Influence Of Demographic Variables onCustomer Satisfaction on E-Banking In Public Sector Banks, Ilkogretim Online - Elementary Education Online, Vol. 20, Issue 5, pp. 1774-1781.

6. Ameme, B. K., (2015), The Impact of Customer Demographic Variables on the Adoption and Use of Internet Banking in

Developing Economies, Journal of Internet Banking and Commerce, Vol. 20, Issue 2.

7. Onyia, P. O., and Tagg, S. K., (2011), Effects of demographic factorson bank customers' attitudes andintention toward Internet bankingadoption in a major developingAfrican country, Journal of Financial Services Marketing.

8. Hill, K., (2004), Study on demographics Vs Marketing' CRM daily

    (online)Available:http://crmdaily.newsfactor.com/story.xhtml?stotytitleWhich_Comes_FirstDemographics_or_Marketing_&story_id=23250&category=lylt,Accessed13Nov.2023

9. Sohail, M., and Shanmugham, B. (2004), E-banking and customers' preferences in Malaysia: an empirical investigation, Information sciences, informatics and Computer Science: an international Journal, Vol. 150, Issue 3-4, pp. 207-217.

6. Sustainable Practice in Finance and Economics.

Ashoka A S

Assistant Professor, GFGC K R puram, Bangalore, Karnataka, **India**

Abstract: This paper explores new theories of sustainable finance. The need for brand spanking new theories of sustainable finance arises from the need to set up a fixed of proposal that could assist us recognise the behavioural movement of economic agents in the supervision or conduct of sustainable finance. The important Six theories of sustainable finance have been explored, namely, the concern theory, idea of sustainable finance, the peer emulation, the life span theory, the nice signalling concept, and the system disruption idea. Those theories provide believable explanations for the behavioural movement of economic marketers towards sustainable finance. Students, teachers, researchers, policy makers, economist will discover these theories very beneficial in their paintings in sustainable finance.

Keywords: Sustainable Finance, theories of sustainable finance,

Introduction

Sustainability is the ability of a population, a communications system of institutions, or a way of life to endure over time. Sustainability is quite often understood from the perspective of inter-generational ethics where capabilities are not diminished by the current generation's Economic and Environmental choices.

The idea of sustainability emerged from the modern Environmental movement, which challenged the unsustainable changing Nature of Civilizations, where resource usage, population growth, and Consumption patterns endanger Ecosystem Integrity and the welfare of future generations. One is presented with sustainability as an Economic solution to ephemeral, confined, and wasteful actions.

It can serve as a Benchmark for Evaluating Organizations and as a target for Society to aim for. In order to improve the performance, the of more sustainable initiatives, sustainability also necessarily requires a critical analysis of the ways which also existing Social structures support cognitive research towards sustainability

Methods to be Sustainable.

The concept of sustainability forms the foundation for sustainable development, sustainable society, and sustainable Production. Sustainable yield involves the extraction of a specific (self-renewing) Natural resource, such as fish or lumber. Such a yield can, in theory, be sustained for ever and ever because the underlying natural system has the regenerative capacity itself. A sustainable society is one that has understood sustaining within Ecological Constraints. It can continue as a whole because processes that hold an excessive amount of strain on the Environment have been Modified or Eliminated.

The practice of advancing society in a way that meets the needs of today's and tomorrow's Citizens, taking into account Economic, Social, and Environmental factors when making Decisions, on the other hand, is more regularly Associated with Environmental limitations or Environmental sustainability.

On results, and this is Frequently Emphasized by the use of the term "Environmental sustainability." Other Common Synonyms for sustainability include "Social sustainability,""Economic sustainability," and "Cultural sustainability," which all refer to long-term threats to welfare in these fields. The concept of "local sustainability" also highlights the importance of "place." Corporate sustainability is another common synonym, and it refers to both the "survival" of a specialized corporation and hence the the"Impact" that Conglomerates can have on the sustainability agenda. crop Central refers to Financial results and Governance Effectiveness. The term "Environment" refers to the idea of "Corporate Responsibility" or "Social Performance," and the term "Corp" is derived from "Corporation." The phrase 'Triple Bottom Line' and the term "Social Performance" are regularly used together to highlight how Important it is for Corporations to pay to attention to Social performance. The word "sustainable" is also used in discussions of Corporate Responsibility, Corporate Reform, and the Creation of Ethical, Environmentally Friendly, and Sustainable Investment Vehicles. **Definition of Sustainable Finance**

The United Nations Environment Programme defines Sustainable Finance as finance that meets the long-term needs of a sustainable and inclusive Economy along all dimensions relevant to achieving those needs, including Economic, Social, and Environmental Issues (UNEP 2017) International Capital Market Association defines Sustainable Finance as thatI incorporates Climate, Green and Social Finance while also adding wider considerations concerning the longer-term Economic sustainability of the Organizations that are being Funded, as well as the Role and Stability of the overall financial system in which they operate (ICMA 2020).

Literature review

Literature evaluation within the larger body of literature on finance, the field of sustainable finance is expanding. According to Migliorelli (2021), there are too many different definitions, concepts, and economic and policy guidelines that control the sustainable finance landscape. Financial institutions must prioritize sustainable funding, according to studies already published in the literature on sustainable finance. For example, Oman and Svartzman (2021) contend that in order to meet the ambitious goal of tackling climate change based on the Paris agreement, policy makers around the world have begun creating sustainable finance programs. They also contend that the financial sector needs to play a greater role in the transition towards a sustainable economy. Shabb, Curtis, and Libertson (2021) note that by integrating sustainability into their financial analysis and portfolios of investments, the finance sector is contributing more and more to the shift to sustainable development. According to Schoenmaker (2018), in order to provide long-term value for the community at large, some financial institutions are starting to steer clear of unsustainable businesses and initiatives. The significance of sustainable finance and investment in Japan is examined by Schumacher, Chenet, and Volz (2020). They additionally discuss how the Japanese financial industry can promote Japan's transition to a zero-carbon, sustainable economy by reducing growing climate risks. They demonstrate the substantial climatic risks to which the Japanese financial sector and its institutions are exposed. Managing Global Transitions: Sustainable Finance Theories 9. acting from both inside and outside of Japan, and the Japanese financial sector has begun to take climate-related risks into account and align itself with the Paris Climate Agreement's 2°C warming scenarios and sustainable development targets. The integration of sustainability in investment banks' underwriting services is examined by Urban and Wójcik (2019). They discover that investment banks refrain from underwriting businesses that supply divisive goods like tobacco, coal, and nuclear weapons, but they nevertheless support underwriters who have been reported for serious environmental, social, and governance misbehavior. The current approach to maximizing shareholder wealth, according to Fatemi and Fooladi (2013), is no longer a reliable framework for establishing sustainable wealth because it places a strong emphasis on short-termism, which has the unintended consequence of forcing many businesses to externalize their social and environmental costs. The creator calls for a move to maintainable fund. Ryszawska (2016) moreover appears that the part of back is changing from the overwhelming see of maximizing benefits and shareholder riches towards one supporting feasible advancement, a green economy, a moo carbon economy, and relief of climate alter. Ozili (2021) proffers arrangements that can make fund ended up maintainable. Ozili (2021) contends that (i) there ought to be more prominent center on how a few viewpoints of fund can contribute to maintainability; (ii) light-touch direction may be required to develop the moderately little feasible back segment; (iii) there's a have to be receive a bottom-up approach to develop the economical fund division; (iv) deliberate esg divulgences and related supportability detailing ought to be energized; and (v) short-term budgetary disobedient can complement long-term rebellious in economical financing. **Sustainable Finance Theories; Some** Sustainable Finance Theories are Presented in this Section. The Theories include the System Disruption Theory of Sustainable Finance, the Positive Signaling Theory of Sustainable Finance, the Life Span Theory of Sustainable Finance, the Peer Emulation Theory of Sustainable Finance, and the Priority theory of Sustainable Finance. **Priority Theory of Sustainable Finance:** According to the Priority theory of sustainable finance, a Country's or Region's Economic Agents' rate of effort toward achieving sustainable finance goals is a true Indicator of how high the sustainable finance agenda is regarded there (Wilson 2010). Three Factors can be used to Evaluate the Priority: (i) the Coordinated, Independent, and Collaborative Efforts made by Economic Agents to achieve sustainable finance goals; (ii) the Speed at which a Consensus is reached; and (iii) the Speed at which actions are taken to achieve those goals. Prioritizing Sustainable Finance Objectives is Necessary as they may take into Account that achieving sustainable finance goals has Consequences. This is due to the Possibility that Prioritizing the achievement of sustainable finance goals may cause other Significant objectives to be neglected until the sustainable finance goals are met. This implies that something will have to be given up in order to Accomplish another aim. These trade-offs can be quite expensive, which makes it inappropriate to give sustainable financial aims Precedence over other Crucial Objectives. For instance, Emerging Nations with pressing demands for Economic Development—like raising Per Capital—might not consider this to be a lesser necessity. The Priority Theory of Sustainable Finance implies that sustainable Financing should be prioritized. **Peer Emulation Theory of Sustainable Finance:** According order to accomplish the Peer Emulation Theory of Sustainable Finance, Economic Agents simulate the Behavioural Patterns, Policies, and Strategies of one's Peers in possess Sustainable Financial Objectives. According to the Peer Emulation Theory of Sustainable Finance, Economic Agents Regress on Adopting similar policies or actions taken by peers they Admire or Emulate when there are no Uniform Standards to Guide Action towards sustainable financing. This implies that Economic Actors will pursue particular sustainable finance goals because their influences already do so or have in the past. There are two Drawbacks to the Peer Emulation Theory of sustainable finance. One is that adopting Peers' sustainable finance policies and practices avoids the differentiated creativity required to develop a new course of action, policy, or strategy from scratch as well as the opportunities for valuable insights. Second, due to differences in financial markets, financial regulation, governance mechanisms, and political will to achieve sustainable finance goals, adopting the same or similar sustainable finance policies and actions that have been adopted by peers in other countries may not produce the desired results.

The Sustainable Finance Life Span Theory: The product cycle hypothesis by Vernon (Vernon 1979) served as the basis for this theory. According to the life span theory of sustainable finance, consumer interest in sustainable financial products, services, instruments, plans, policies, or activities is influenced by how long they last. It makes a strong case the that impartial financial intermediaries are aware that sustainable finance products, services, instruments, schemes, policies, or activities (hereinafter, "sustainable finance products") have a life cycle that starts with the introduction of sustainable finance as a new concept and ends with its decline. Economic agents can decide whether to make a short-term commitment, a long-term commitment, or no commitment at all to enduring contribution finance based on their knowledge of the life cycle of finance products, which enables them to independently predict the estimated life span of finance products. This means that the perceived life span of a particular sustainable finance product by economic agents determines the extent of their support for sustainable finance and the extent to toward which they support the transition from traditional/mainstream financing to sustainable financing. **The Sustainable Finance Theory of System Disruption**. According to the sustainable finance theory of system disruption, pursuing sustainable finance objectives may cause major disruption in the traditional/mainstream financial system along with disruption of businesses that heavily rely on traditional/mainstream financing. Depending on the degree of the disruption brought on by the development of sustainable-to-sustainable finance, the influenced economic agents may either resist the change or decide not to support the sustainable finance initiative. According to this theory, economic agents may be forced to choose whether or not to support or take an active part in the transition to sustainable finance due to the potential disruption to the existing system (traditional/mainstream finance) affected by it. Economic agents will base their choice on whether they believe that using sustainable finance will be more advantageous than it will be expensive and whether the subsequent disruption will have a significant impact on their business, income, or way of life.

Positive Signalling Theory: Positive signalling theory posits that economic agents have an incentive to disclose positive information about their commitment to pursue one or more sustainable financial goals in order to signal good news to external parties who have can support their goals (Quatrini 2021; Park 2018), Economic agents can disclose positive information about their sustainable financial intentions by publishing directly in the media or by voluntarily providing additional financial and non-financial information. in its published annual reports. For example, companies can publish information about their latest sustainable finance instruments or green bonds to attract investors who want to invest money in sustainability-oriented companies. Such a disclosure makes it easier to attract investors interested in green bonds. Likewise, the government could publicly announce that it will announce a national sustainable finance policy. Such an announcement would not only strengthen the country's reputation for sustainability but could also signal that the country is willing to receive foreign technical assistance as it implements its national sustainable finance policy and has can attract huge foreign direct investment aimed at green projects in the region.

Sustainable Finance Resource Theory: According to the resource theory of sustainable finance, some countries have improved markedly more than others in achieving their sustainable finance goals as a result of differences in the human-made resources that can support those goals. According to the resource theory of sustainable finance, some regions have superior man-made resources that give them a comparative advantage over other in achieving their sustainable finance objectives and achieving the relay to sustainable finance. For instance, some nations have a large amount of foreign reserves, a budget surplus, low external debt, a developed financial sector, splitting financial technology systems, strict financial regulation and supervision, performance as compared for monitoring climate change, better sustainability education, a population that is environmental conscious, and a significant number of institutional investors willing to invest in sustainable finance instruments. Countries that have direct connections to these innumerable man-made resources have a comparative advantage and can, as a result, perform their tasks for sustainable finance more quickly and easily than nations that do not. Relatively to nations with a no availability of foreign reserves, a sizable budget deficit, a high level of external debt, an underdeveloped financial sector, inadequate financial technology systems, permissive financial regulation and supervision, and few or no institutional investors willing to invest in sustainable finance instruments, countries with an abundant supply of man-made resources can also make the transition from traditional/mainstream finance to sustainable finance extra quickly Sustainability, Commerce And Economics: How To Create Right Balance For Sustainable Development. We may be losing the war against climate change, as we fail to approach sustainability-related issues in the right manner. In a recent article 'The Invisible Hand Won't Solve the Climate Crisis. Capitalism Must Evolve', US professor Andrew J. Hoffman well articulates the perils of global unsustainability and the need to redefine the role of corporation and market. The Economist magazine also warns us against the dire consequence of not addressing the sustainability and climate change related issues in its August month's cover story 'In the line of fire - losing the war against climate change'. Both the articles represent a view that is increasingly gaining strength - a view that suggests that despite all the tall talks over sustainable development, we may be losing the war against climate change, as we fail to approach sustainability-related issues in the right manner. The reason is that all the standards - in global trade, in bilateral negotiations, industrial and investment policies and the framework for private and inter-governmental bodies - that we have been setting all the while, have been production oriented, and not consumer-centre Ever since the United Nations Development Programme (UNDP) set 17 sustainable development goals (SDGs) in 2015 at a meeting in Paris, all economic and developmental activities post-2015 are supposed to have the sustainability agenda, with an aim to protect all living creatures inclusive of flora and fauna, at its core. A mechanism to tackle climate change issues are also supposed to be in-built. However, in practice, there is an imbalance in the process of new standards (public policy measures) that are, in principle, guided by the SDGs. The non-mandatory standards set by the United Nations Forum for Many other activities aggravate the imbalance between consumption and production. For instance, the standards notified by the World Trade Organisation (WTO) members are also framed after extensive consultations with the private sector. According to an internal analysis done by the Centre for WTO Studies of the Indian Institute of Foreign Trade (IIFT), WTO's present classification suggests that technical barriers to trade (TBT) notifications by the developing countries are higher when compared to the developed and the least developed countries (LDCs). The situation is diametrically opposite when income or industry-based classifications, provided by the World Bank and United Nations Industrial Development Organisation (UNIDO) respectively, are considered. The higher income countries and countries with high GDP shares of the manufacturing sector are dominant users of TBT standards. Also, the developed countries TBT notifications have a higher weight for the private sector, which regulates production activities. The trade policymaking in the industrial countries thus has marked a shift in the direction that favours more and more public policy measures as a trade policy instrument. This has led to a proliferation of standards while the tariffs continued to lose its relevance as a barrier to international trade. The policy shift facilitated the creation of non-tariff measures in the areas of namely environment (sustainability and climate change), labour and others technical standards. Of the 17 SDGs, Goal 12, which talks about "Responsible consumption and production", attempts to address sustainability issues holistically by having both consumption and production related standards/controls. It states, "Achieving economic growth and sustainable development requires that we urgently reduce our ecological footprint by changing the way we development, we need to keep a check on three fundamental pillars: social progress, economic development, environment, and climate. Therefore, it will become essential for economic and commercial activity and practices to be compliant with sustainability standards based on production and consumption. Since the word sustainability is expressed in three different ways: eco labels, messages and claims about a product or process, the standards are to be developed with the need to have sustainability incorporated in all such activities of procurement, production and trading. Of the 500 plus private standards that we have today, almost all are production-regulating standards. There are no standards or capping the consumption of various goods, except for a few standards regulating the marine fish stocks for intermediate products and as labels, marks and certificates for consumer's information. A study conducted by the Food and Agricultural Organisation (FAO) of the United Nations (UN) is a pointer to the inadequacy of the consumption-based standard as it tells us about the quantum of waste generated by the big multinational food retail stores. It says that the per capita waste by consumers in Europe and North America is between 95-115 kg compared to 6-11 kg of those in Sub-Saharan Africa, South and South-Eastern Asia.

Table 1: Sustainability Development

Element	Criteria
Economic stability	Growth Development Productivity Trickle Down
Social sustainability	Equity Empowerment Accessibility Participation Sharing Cultural Identity Institutional Stability
Environmental Sustainability	Eco-System Integrity Carrying Capacity Biodiversity

Figure 2: Sustainability Practices in Commerce and Economics

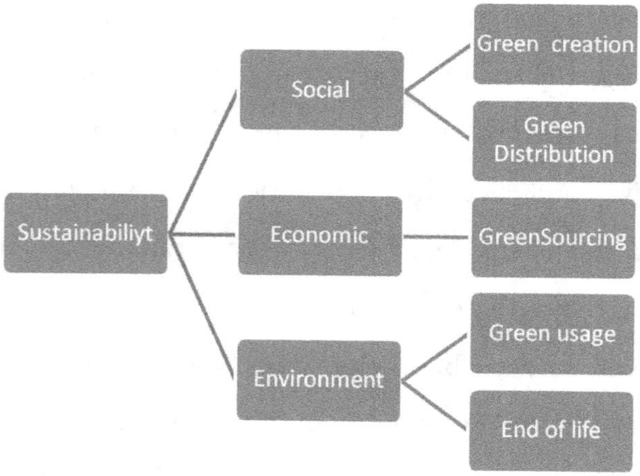

Secondary source; AIP Conference Proceedings 2129(1):020167

Conclusion: This study developed several sustainable finance ideas to promote the sustainability discussion in academic and business communities. The peer emulation theory, the resource theory, and the priority theory of sustainable finance were all developed. the system disruption theory of sustainable finance, the life span theory of sustainable finance, the positive signalling theory of sustainable finance, and others. These ideas provide explanations for the actions and choices taken by economic actors in assistance of the sustainable financial agenda. These theories might indeed combine to position at the moment discussion threads taking about sustainable finance, which is still a developing field of study. The theories of sustainable finance that have been developed have implications for developing a strong foundation to comprehend the actions and conduct of economic agents with regard to sustainable finance. Improvements in sustainable financing could present novel challenges and opportunities for theory development, opening up new avenues for research.

References

1. Ahlström, H., and D. Monciardini. 2021. 'The Regulatory Dynamics of Sustainable Finance: Paradoxical Success and Limitations of eu Reforms.' Journal of Business Ethics 177:193–212.

2. Shabb, K., S. Curtis, and F. Libertson. 2021. 'Sustainable Finance: Investing in Our Future.' Lund University. https://soundcloud.com/iiieepodcast /sustainable-finance.

3. Shalneva, M. S., and Y. V. Zinchenko. 2018. 'Sustainable Finance as a Way of European Companies' Transition to Green Economy.' In International Conference Project "The future of the Global Financial System:

4. Arouri, M., El Ghoul, S., & Gomes, M. (2021). Greenwashing and product market competition. *Finance Research Letters, 42,* 101927. (in press).

5. M., &Mussalli, G. (2020). An integrated approach to quantitative ESG investing. *The Journal of Portfolio Management, 46*(3), 65–

6. D. BASIAGO,(1999) Economic, social, and environmental sustainability in development theory and urban planning

practice, The Environmentalist vol 19, pp 145-161 . 9 Kluwer Academic Publishers, Boston. Manufactured in the Netherlands.

7. Downfall of Harmony," edited by G. Popkova, 1002–12. Cham: Springer. Sommer, S. 2020. 'Sustainable Finance: An Overview.' Deutsche Gesellschaftfür Internationale Zusammenarbeit (g iz) GmbH, Bonn; EschborngizAgència Brasília scn, Brasília.

8. POJK.03/2017 (2017) 3. D. Cash, SSRN (July 2017

9. Ozili, Peterson K, Theories of Sustainable Finance (March 1, 2022). Managing Global Transitions, March 2022.

10. https://www.businesstoday.in/opinion/columns/story/sustainability-commerce-and-economics-how-to-create-right-balance-for-sustainable-development-151383-2018-10-17

7. Towards the Development of Digital Competencies Assessment Model (DigiCAM) on Distance Learning of Secondary Schools in Antipolo City.

Maricel L. Gebilaguin

Far Eastern University, Dep. Ed-Bagong Nayon II National High School, **Philippines**

Abstract: Digital assessment is the use of technology to develop, deliver, score, and analyze examinations. Summative tests and performance tasks were given through electronic devices such as personal computers, laptops and mobile phoneswith the use of internet. Nowadays, teachers in public schools are embracing the use of different digital platforms in giving summative assessment to the students in distance learning mode. The use of different digital platforms can help teachers administer summative assessment in an online distance learning efficiently. Students in this time of pandemic hardly submittheir summative assessment on time due to several reasons such ashelping in family chores, working, and overlapping activities. Thus, to alleviate teachers' work in checking, and collecting late summative assessments, a qualitative case study has been done to one of the schools in Antipolo City: the Antipolo Senior High School. It was the first school in the division of Antipolo to achieve an advanced category for the SBM level of practice which was validated by the DepEdRegion IV-A CALABARZON. This study explored the challenges and opportunities in conducting summative assessment using different application platforms suited to the capability of the students. 15 participants were selected purposively with certain criteria.They participated through focus group discussion and individual interview. The researcher used the triangulation method like artifacts, interview, FGD, and reviewed articles. The developed Digital Competencies Assessment Model (DigiCAM) would help other schools to have an idea on how to improve their MPS Result as well as their SBM level of practice.

Keywords: digital assessment, SBM level of practice, summative assessment

Introduction

In distance learning modality, teachers made printed summative assessments with the students' modules. These materials were collected by the parents from the school where their child enrolled. At the end of the grading period,summative assessment should be administered to measure the degree of learners' mastery on the most essential learning competencies as stipulated in DepEd order No. 031 s. 2020. Thus, the result of this assessment tool is recorded and is used as a basis of learners' achievement in every learning area. Digital assessment supports distance learning where students can access anywhere. Today, our education is highly influenced by technology and to enhance students' academic achievement schools are expected to use it. Today, our education is highly influenced by technology and schools must utilize such tool to attain remarkable increase on students' academic achievement.

Literature Review

The development of Digital Competencies Assessment Model (DigiCAM) explored the challenges and opportunities of teachers using digital assessment in distance learning. This pandemic, educational sectors all over the world formulate on how to consider teaching and learning through meaningful assessment practices (Scull et al.,2020). The occurrence of pandemic brought eagerness to the global educational sectors to think of ways on how to asses teaching and learning with the use of purposeful assessment practices. It comprehends how teachers assess students' progress on both formative and summative assessment; how teachers distribute graded activities across the learning area; the issues involved providing effective feedback, and the strategies in distance learning modality. **Teachers' Digital Competencies Teachers** need to have an awareness on factors that impact their use of technology in teaching such as their opinions, and attitudes. According to Winter et al. (2021), some of the teachers have negative attitudes toward technology because they do not see its usefulness to their discipline. According to Winter et al. (2021), some teachers perceived the use of technology negatively as they fail to realize its worth and usefulness to their profession. As a teacher, being innovative in this new normal is one of the most important things to do especially in the basic education sector. But, how do teachers admit the negative attitudes toward technology? Technology has become part of teaching "Teachnology". Exploring technology will give us more understanding of 21st century competencies. However, students in public schools do not have the resources for distant learning. Those students who are in the digitized learning modality have phones to access the soft copies of modules. They also have options for the submission of their written work: online or on a piece of paper. Generally, teachers have gadgets to be used to work online but must be explored and utilized with different apps despite of its high expenditure. They must explore and learn different apps that can be used for online assessment for this would make their work easier than the usual. Knowing the challenges and opportunities in dealing with the digital assessment challenges and effective practices of Antipolo Senior High School that acquired advanced level in SBM level of practice as validated by the Region IV-A office S.Y. 2020-2021 may help to illuminate the next steps in developing a model for digital competencies in assessing students in distance learning.

Therefore, this study pursued to answer following research questions:

1. What are the digital challenges and opportunities of teachers in giving assessment on distance learning modality?

2. How do the teachers address the challenges in online assessment?

3. How does the Digital Competencies Assessment Model (DigiCAM) give opportunities to teachers to increase the performance rating of their students in distance learning?

Scope and Delimitation This study was conducted to a school that achieved an advanced SBM level of practice in secondary school of Antipolo City on S.Y. 2020-2021. The researcher sent a letter to the division of Antipolo to know the names of the school/s that belong/s to the SBM advanced level category. Based on the data released by the Division of Antipolo, only one secondary school has an advanced level of practice and that is Antipolo City Senior High School (ACSHS). It is the only stand-alone Senior high School in the Division of Antipolo City. It was established in 2016 through Republic Act 10533 otherwise known as the Enhanced Basic Education Act of 2013 which requires two years of Senior High School in Basic Education.

Methodology

This study was guided by an open-ended question to address the challenges and opportunities of teachers in giving summative assessment on distance learning. An interview protocol was made and forwarded to the participants before the interview. The researcher used purposive sampling because it's a technique used widely in qualitative research in identifying and selecting information-rich cases. It is a popular method used by the researchers because of its extremely time and cost effective (Alchemer,2020). For this study, the researcher utilized purposive sampling. It is the most used technique in determining and selecting cases that are rich in information because of its efficiency and good results with a low cost (Alchemer,2020). The researcher identified and selected individuals as respondents who met the given criteria. The highly proficient and proficient teachers were very knowledgeable and experienced in making digital assessment. In fact, the ACSHS reached average scores, nearly in performance rating.

Cases category	Position
Case 1: Master Teacher	9 MT's (3 Science majors, 2 Math majors 1 social science)
Case 2: Teacher III	3-(1Science major, 2-English major coordinator (2) key teacher (1)
Case 3: Teacher I-II	3 (STEM, coordinator (1), key teacher (2)

The participants of this study were nine with a combination of Master Teachers I & II, three coordinators and three key teachers of their departments. Six of them are Science majors, three are Math majors and three Social Science majors. All of them are teaching STEM, and HUMMS at Antipolo City Senior High School. There were fifteen respondents but unfortunately, the researcher failed to conduct the follow up interview due to time contains and unavailability of the respondents. There were fifteen (15) participantsinvolved in this study and categorized as Master teacher (case 1), Teacher III (case 2), and Teacher I-II (case 3). Each case has three or more sub unit analysis. On Case 1, nine MT's have been analyzed: three teachers for case 2, and three teachers for case 3. The cross-case analysis (table 4) was used in this data. It's a drawing on challenges and opportunities of teachers in giving online assessment in distance learning. As shown in the study of Cruzes et al,.2014, and cited in Schwarbach&Onoh, (2021), cross case analysis is a method that is used to depict the likeness and differences of each case with respect to the occurrences, activities and processes. Explained cross case analysis as a technique that aids the comparison of what a unit of a case has in common and the difference in the occurrences, activities and processes. This technique was used by the researcher to compare the participants' experiences in giving online assessments. Hence, this technique revealed that some old teachers were struggling in dealing with technology and it is one of the major factors influencing our education not only in the Philippines but all over the world. The researchers followed the procedural framework in Figure 4

For Step 1: *Develop Theory*: The researchers developed the research questions that were used during the conduct of the interview and some follow-up questions that came-up with the in-depth answers of the respondents For Step 2: *Designing Data Collection Protocol*: An interview protocol was sent to the respondents via google form before the conduct of the interview. For Step 3. *Select Cases*: After the respondents agreed in the given interview protocol, an interview happened depending on the preferred platform of the respondent. Interview were conducted thru face-to-face for FGD and via Zoom. For Step 4: *Conduct Cases Study*: After the transcriptions, three cases with each sub unit. (Case 1 – 6-Master teachers, Case 2 – Teacher II-III (3), Case 3 –3 Teachers I with 10 years in service and acted as coordinator and key teacher in their department. Three major themes and six sub themes were identified and developed and from these, the Digital Competencies Assessment Model (DigiCAM) was formed.

For Step 5: *Write Individual Case Reports*: For the final step, an individual case report was written in each case based on the themes which answered the research questions used in this study. Each case report describes teachers' challenges and opportunities in giving summative assessment on distance learning.

Figure 1 *Procedural Framework of this Multiple-Case Study (Yin, 2014)*

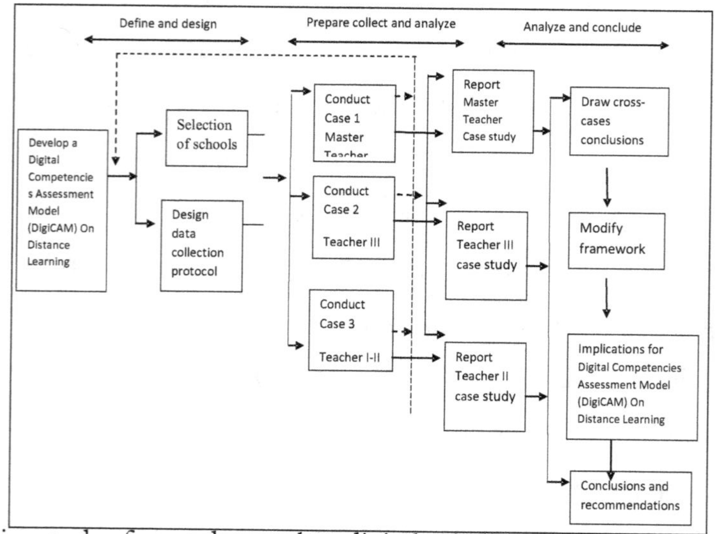

This study focused on the digital challenges and opportunities of teachers in different learning areas through modular distance learning. The participants were selected purposivelyand interviewed by the researcher individually. Out of 16 chosen participants, 6 only participated in focus group discussions. Qualitative sampling approaches were used, which means that the sample population was drawn and from those sampling, representatives were taken. The data were collected to explore the unit of analysis, which may be a phenomenon or lived experience (Dieumegard et al., 2019). The data were gathered and used in investigating the phenomenon and lived experiences of the participants. The purposeful sampling was used to ask participants to identify specific subjects that could help to answer the questions. And it allows the researcher to focus in depth on a phenomenon (Schoch,2020). The researcher actively selected the most productive sample to answer the research question. This study is a qualitative case study to be specific. In addition, to make the data rich and reach the "saturation" stage of gathering data, focus group discussion was used by the researcher. Focus group discussions or FGD's was also used because it is useful in exploring, developing and refining ideas, gathering initial research questions as well as in making interview schedules. It is also one the most common tools used for qualitative research (Vanderstoep & Johnston, 2009, as cited inTümen-Akyıldız, S. & Ahmed, K.H,2021). This study aimed to develop digital assessment model from a focus group discussion, personal interview and artifacts from the participants to gather rich data. Thus, a collective set of values, experiences, and observations of participants will be achieved also. Ethical Issues The respondents were the highly proficient and proficient teachers of Antipolo Senior High School — the only stand-alone senior high school in the division of Antipolo and the only secondary school that reached the advanced SBM level of practice which was validated by the Region IV-A. The researcher established trust with the selected respondents through ensuring the anonymity and confidentiality of all respondents. The researcher carefully explained the process on how the data will be presented. Also, the researcher provided more information about the project as well as its goal without influencing the responses. The data were gathered through interviews and focus group discussions. The researcher provided a letter of acceptance beforehand for the time and date of the specific interviews according to the own convenience of the participants. Identification of ethical concerns that could possibly emanate from the conduct of research, and discussion on how to prevent these from taking place were also put in consideration. Lastly, the researcher ensured the rightful way of conducting a study or investigation to answer the given questions. She also secured free prior and informed consent from respondents and issued confidentiality and anonymity.

Results

Three major themes and six sub themes were developed in giving digital assessment of teachers on distance learning.

Figure 5 Themes & sub themes

Research Questions	Major themes	Sub themes
1. What are the digital challenges and	Digital readiness	-Use accessible friendly and interactive apps -Use effective digital

		communication tool -Explore innovation and utilize technology
opportunities of teachers in giving assessment on distance learning modality?		
2. How do the teachers address the challenges in online assessment?	Parents/Guardians Involvement	-Open Communication
3. How does the digital Competencies Assessment Model (DigiCAM) give opportunity to teacher's increase performance	Digital feedback	-Tasks made easier with technology -Allow second chances

rating on distance learning?		

Theme 1: Digital readiness

Teachers 'digital readiness depends on how students and parents respond to teacher-provided tasks. The eventual goal is to streamline the way teaching is done, and learners are served or vice versa, Martin, (2021). Teachers nowadays are embracing the use of different platforms in giving assessment. *Ms R said that "I only used to google forms in the google classroom. I don't want to use any application, kasi as of now ito lang talagah ang platform naalam ng karamihan ng mgabata. Basically, I used to google forms and the link will be uploaded in the goggle classroom kasi easy access po itosamgabata."* Being digitally ready means exploring different applications that we can use in assessing our students. By using different apps, we could assess our students in modular or online distance learning in an easy way. Assessments can motivate learners to have goals and at the same time, it can evaluate teaching effectiveness. One of the respondents, Mr. A said, *"The best assessment in digital form that I usually practice from last year up to now is the quizizz which is a gamified assessment learning website. It is an exciting tool for learning in giving assessment because there's a lot of questions that I can give to my learners and assign from that there is an entertainment while learning."* With the use of accessible, friendly, and interactive apps, learners can be motivated to learn. Teachers can promote enthusiasm and enjoyment by capturing the interest of learners and eventually inspire learners also to pursue learning despite of hardship in life. In addition, teachers can use different devices for effective digital communication as a tool in reaching the students such as; mobiles phones and laptops. As Mr. P mentioned, *"mobile phones are the best tool to communicate with students especially in giving feed backs"*. Mr P. said that *"through call and text messages, students were easily reached out"*. He added that *"most of the time, students have no load to respond to teachers immediately"*. Ms. R agreed with what Mr. P *said*. Ms. R explained, *"I also used messenger in FB, this also another tool I used to have an effective communication to my students."* For her, messenger is also one of the best apps in reaching her studentsbecause her students could still see her messages even if they have no load to see the files. Explore innovation and utilize technology. Innovation refers to something new and different from before as defined by the dictionary.com. As (AECT, n.d.as cited in Bui, 2020) stated that *"facilitating learning and improving performance by creating, using, and managing appropriate technological processes and resources."* Teachers need to explore different innovations in teaching. Innovation is also often considered as a new product. For instance, to use a gamified apps in assessing our students can be said as new because it is different from the usual activity that the teacher does in giving assessment. As Mrs. Z said, "Although *I'm not young but I explore different apps, I used different gadgets in giving assessment to my students, ayaw ko naman namapag-iiwananako. Everything can be learned. Our generation nowadays are inclined to different gadgets such as android phones computers tablets etc."* Mrs. Z also explained that students actively participated in social communication such as

Facebook and messenger since early childhood that's why they were not active in answering their modules. "As *a teacher of this new generations I need to explore innovate and utilized technologies because that's the trend nowadays"* Mrs Z. added.

Dissemination and Advocacy Plans

1. The result of this study will be shared with Teachers during LAC sessions, so that they may have an awareness on the Digital Competencies Assessment Model (DigiCAM) on distance learning). Although now is already a face-to-face mode of learning, proper communication with students is a must.

2. The DigiCAM Model can help parents know that their children need help in their studies with proper communication with their teachers

References

1. Beauchamp., Moskeland, A., Milner-Gulland, E.J., Hirons, M., Ruli, B., Byg, A.,Dougill,A., Jew, E., Keane, A.,Malhi, Y., McNicol, I., Morel, A., Whitefield, S., &Morris, R. (2019). *The role of quantitative cross-case analysis in understanding Tropical smallholder farmers' adaptive capacity to climate shocks.* Environmental Research, 14,125013. https://doi.org/10.1088/1748-9326/ab59c8

2. DepEd Order No. 031 s. (2020). *Interim Guidelines for Assessment and Gradingin Light of the Basic Education Learning Continuity Plan.*

3. Gumapac., Aytona E., Alba, M. (2021). *Parents Involvement in AccomplishingStudents Learning Tasks in the New Normal.* International Journal of Research inEngineering, Science and Management,4(7) ,2581-5792.

4. Gustafsson, J. (2017). *Single case studies vs. multiple case studies:A comparative study.*

5. Hodges, C. B., & T. A. Cullen. (2020). *Teacher Readiness to Implement TechnologyInnovations.* i-Manager's School Educational Technology, 16 (2), 1–8.https://doi.org/10.26634/jsch.16.2.17539

6. Liberman,J., Levin, V,. &Bazaldua,D.L.(2020).Are students still learning during COVID-19? F Formative assessment can provide the answer.https://blogs.worldbank.org/education

7. Scheelbeek, P., Hamza, Y., Schellenberg, J., & Hill, Z. (2020). *Improving the use offocus group discussions in low income settings.* BMC Medical Research Methodology, 20, 287. https://doi.org/10.1186/s12874-020-01168-8

8. Schlichter, A. 2020. *The Impact of Covid-19 on Education: Insights from Educationat A Glance.* Paris: OECD Publishing.

9. Sarmiento, C.P., Morales, M.P.E., Elipane, L.E., & Palomar, B.C. 2020). *Assessment practices in Philippine higher STEAM education.* Journal of University Teaching & Learning Practice,17(5). https://ro.uow.edu.au/jutlp/vol17/iss5/18

10. Ribeiro,R. (2019). *Digital and collaborative feedback: promoting students' awareness of their learning.*

11. https://www.cambridge.org/elt/blog/2019/12/09/digital-and-collaborative-feedback/

12. Taylor, S.P. (2017) *What Is Innovation? A Study of the Definitions, Academic Modelsand Applicability of Innovation to an Example of Social Housing in England.* OpenJournal of Social Sciences, 5, 128-146.https://doi.org/10.4236/jss.2017.511010

13. Tümen-Akyıldız, S., & Ahmed, K.H. (2021). *An overview of qualitative research andfocus group discussion.* Journal of Academic Research in Education, 7(1), 1-15.https://doi.org/10.17985/ijare.866762

14. Winter, E., Costello, A., O'Brien, M., & Hickey, G. (2021). *Teachers' use of technology and the impact of Covid-19.* Irish Educational Studies, 40 (2), 235-246.https://doi.org/10.1080/03323315.2021.1916559

15. Yin, R.K. (2018). *Case study research: Design and methods* (6th ed.). Thousandoaks, CA: SAGE.

8. Contours of Gandhian Concept.

Dr. Saraswathi. K

Associate Professor Department of Post-Graduation Studies in Political Science, Government First Grade College, K.R.Puram, Bangalore, Karnataka, **India**.

Abstract: Great Humanist, Freedom Fighter, Truth, Non-Violence, simplicity, tolerance, Sarvodaya, Satyagraha, 'My life is my message' - The great icon who proclaimed to the world, M.K. Gandhi known as 'Father of the nation', who gave the world inspiration, ideas and powerful strategies for tackling violence through the weapon of Non-Violence in a wide range of contexts. Gandhian Concepts are taking different contours throughout the world in a different manner. Arms competition between nations has caused unrest. Covid-19 Coronavirus Disease has brought to the world a reassessment of the concept of Gandhism. His thoughts are further strengthened. As stated in Gandhism, It is essential to follow the principle of cooperation instead of competition. Everyday violence, rape of women, and drunk and drive cases are increasing rampantly, it has ruined people morally, and physically economically and it has destroyed the happiness of the family. Gandhi regarded drinkers as a "diseased man quite unable to help himself".Young people at a very young age become addicted to alcohol. Alcohol acceptance has become a common process at birthday parties, and amusement/entertainment parties. Gandhi protested against alcohol did satyagraha against the government to demand the prohibition of liquor. The number of drunkards in India is increasing rapidly from year to year. Indians are competing in a way that we are no less than Western countries and more and more young people are addicted to drinking. The per capita alcohol consumption in India increased by twofold According to the Global Status Report on Alcohol and Health "Indians consumed 2.4 litre of alcohol in 2005, which increased to 4.3 litre in 2010 and scaled up to 5.7 litre in 2016".

Keywords: Mahatma Gandhi, Indian Youth, Media, Policy, States, women.

Introduction: Young people at a very young age become addicted to alcohol. Alcohol acceptance has become a common process at birthday parties, and amusement/entertainment parties. Gandhi protested against alcohol did satyagraha against the government to demand the prohibition of liquor. The number of drunkards in India is increasing rapidly from year to year. Indians are competing in a way that we are no less than Western countries and more and more young people are addicted to drinking. The per capita alcohol consumption in India increased by twofold According to the Global Status Report on Alcohol and Health "Indians consumed 2.4 litres of alcohol in 2005, which increased to 4.3 litres in 2010 and scaled up to 5.7 litres in 2016".

Review of Literature

Report of World Health Organisation statistics reflects that Indians are becoming drinkers and causing different adversaries on Physical, mental, and social effects on society. Mahatma Gandhi. Young India quoted in David M. Fahey and Padma Manian, Article reveals Protest against alcohol started by Mahatma Gandhi in India. Mahatma Gandhi, The Storey of My Experiments with Truth-An Autobiography Researcher Mahima Nayar'and an Article on alcohol addiction and its effects on Women with a titled interview published in the Journal In the Shadow of Alcohol: Women's Experience of Bangalore Saxena.S, in "Country profile on alcohol in India" stated, how India is becoming the largest consumer of alcohol at the global level.

Methodology: In this study, the historical, analytical, and comparative method has been used to know the role of Mahatma Gandhi against alcohol, Constitutional provisions towards Gandhian Ideology, and women's movement against liquor. Etc.,

Description: According to the World Health Organisation Report of 2020, Alcohol consumption contributes to 3 million deaths each year globally as well as to the disabilities and poor health of millions of people. In 1925, Gandhi argued, "The one thing most deplorable next to untouchability is the drink curse" (All about Travancore", Young

India, 26 March 1925, collected works of Mahatma Gandhi, Delhi publications Division, Ministry of Information and Broadcasting, Government of India, 1958-84) Gandhism is not only relevant today if it follows some of its principles, it definitely not only India will definitely become a superpower country Most of the problems that we are facing today there is the solution in Gandhian Concepts. We can see three categories of followers of Gandhism, among them category one- who just for the namesake Gandhian followers, second, one who use Gandhian Name to fulfil their own selfish attitudes, the last category - one who doesn't preach, but they have practice Gandhian Principles throughout their life. Liquor shops raise major revenue for the government because it is confident that can make a profit by selling liquor to attract customers without any trouble. The number of drunkards in India is increasing rapidly from year to year, more and more young people are addicted to drinking. Gandhi proposed the ban on liquor, which was then protested by some Parsi brewers and merchants and the British government. Excise was the government's biggest income except land revenue. But most of the Indian drinkers are poor and illiterate. Although most of his British friends drank, Gandhi never tried, and before he left India for London, Gandhi swore to his mother, "Don't touch wine, woman and meat" while in England.

The case of Karnataka: Karnataka Government ordered a 14-day "corona curfew" instead of a lockdown announced lockdown imposed on the state from the evening of 27 April 2021(Deccan Herald, 27/04/2021) put restrictions to contain the pandemic. When the statement was released by the Chief Minister immediately people rushed towards the shops to purchase essential things at the same time people rushed towards the liquor shops to purchase liquor Shops and Governments that encourage drinkers have brought tears to wives, sisters, and mothers of sons who have been victims of drunker. Due to the sale of alcohol, poverty, domestic violence, crime and the destruction of many families have resulted. All the victims decided to take the Gandhian route to tackle the problem. In the State of Karnataka Hyderabad Karnataka- the most backward region in the state started an agitation against liquor (The Hindu, 27 January 2019) over Four Thousand women started a movement on 19 January 2019 seeking the demand of ban on Liquor in the state (Madya Nishedh Andolan) taking padayatra (hiking) from Chitradurga to Vidhana Soudha (State Legislature of Karnataka) of Bengaluru with more women were joined from over 20 Districts of the State including Belagavi, Kalaburgi, Ballari, Koppal, Yadgir, Bidar, Haveri, Chitradurga, Raichur, Dharwad, Bagalkot, and Tumkur and demanded complete liquor ban in the state because many women who suffered those who their husbands, children or father's bad drinking habit and leading suffocating life, they were in the group of 20-70 years in the aged and they covered 20 Km per day. The 12-day hiking (padayatra) movement started from Chitradurga and reached 20 km each day to reach Bengaluru. The movement was formed in 2016 to demand the prohibition of liquor in Karnataka in Raichur, where more than 45,000 women had gathered, demanding a liquor ban in the state. Seventy-one-day dharna (to mark 71 years of Independence) sitting of 30 days dharna and gave representation to all major political parties, during that time prohibition could not be implemented in the state and should be a uniform nationwide policy to succeed replied by the then Chief Minister of Karnataka Siddaramaiah. Women had been holding public meetings in the villages they passed through where they created awareness among women about the ill effects of alcohol. The drive was supported by 56 organizations across the state. The protesters' main demand was, a ban on the production, sale, and distribution of liquor, which according to them, due to liquor sale resulting in poverty, domestic Violence, and Crimes and destroyed many families. According to the statement, Prasanna, a Theatre person, "The State's total revenue is Rs. 2 lakh Crore. The Revenue from the Excise Department is Rs. 18.000 Crore. We don't know why the government is giving so much importance to liquor, which is destroying many lives, families, and social health of society". One of the volunteers, Gubbamma from Chincholi in the Raichur district, said "I am 60 years old, but I am participating in the padayatra for the sake of my 20-year-old son. The little money earned to feed hungry stomachs in the house by working as laborers is snatched away for liquor" Manjula from Guduru in Bagalkote district expressed in the interview, "My daughter was widowed within three months after the marriage. Her husband died in an accident when he was in an inebriated state".Statement given by Swarna, a state committee member of the movement, "It is a myth that the government is running on the revenue earned from the Excise

Department. Permitting liquor sales is against the Constitution as the Constitution directs the state to raise the level of nutrition and standard of living and to improve public health as among its primary duties. There are states that are doing well despite the liquor ban. In the eight years, the excise revenue has increased from R.9.000 to Rs 18.000 crore.'don't want liquor, needed water', (Beer Beda, Neer Beku) Karnataka Women march to Bengaluru (28 January 2019) joined women from Mandya, Mysuru, Ramanagara and Chamarajanagar, Tumkur not associated with any political parties demanded the complete prohibition of liquor sales. In the Hyderabad-Karnataka region, alcohol is found in Kirana shops (very small shops) or pan shops apart from bars and wine stores. Being the border district, spurious liquor comes from the other side. Families are in bad condition. Men have sold whatever is available at home, including steel utensils and gadgets to drink. (12 May 2018, The Hindu) March faced many problems in that one of the activists died in a Road accident on NH-48, on March (29, January 2019, Indian Express) 57-year-old Renukamma had died in the road accident when a speeding bike hit her at Nelamangala on NH-48. On night 28 January 2019 hailed from Tairawadagi of Lingasuguru Taluk in Raichur District, "She took an active part in the struggle for three years as her family was affected by the alcohol menace" There is a causal relationship between harmful use of alcohol and a range of mental and behavioral disorders, other noncommunicable conditions as well as injuries, and resultant poverty. One of the initial studies done by Gayford reported that drinking accompanied 44% of cases of women abuse (quoted in Mahima Nayar, In the shadow of alcohol: Women's experience of Bangalore Indian anthropologist, Vol. 39, No. 1/2 (Jan-Dec 2009) PP 49-64 The states of Andhra Pradesh, Telangana, Kerala, Karnataka, Sikkim, Haryana, and Himachal Pradesh are amongst the largest consumers of alcohol in India. (https://www.ambrosiaindia.com/2021/02/4494) Women consuming alcohol has also increased in recent years. Research shows that around half of the adults will consume by 2030. Chief Minister of Karnataka expressed his helplessness to impose the ban on alcohol (29, January 2019, Indian Express) when the activists met to tackle the problem. (12 May 2018) The government opened liquor shops first even though temple mosques were not allowed to open, people rushed towards the liquor shops to purchase liquor, in the first wave of Covid-19. Gandhi says that if I were appointed as a dictator of this country "If I appointed dictator for one hour for all India, the The first thing I would do would be to close without compensation all the liquor shops, destroy all the toddy palms such as I know them in Gujarat, compel factory owners to produce humane conditions for their workmen, and open refreshment and recreation rooms where these workmen would get innocent drinks and equally innocent amusements." (Young India, 25 June 1931, 47:53-54) Gandhi started the Satyagraha movement against liquor. Gandhi was confident that India had the capacity and opportunity to lead the world in liquor prohibition. British politician Caine helped to organize temperance meetings in India against Anti anti-drink movement, Indian National Congress, Arya Samaj, and

Brahma Samaj helped the movement. According to Gandhi Poorer were the greatest sufferers. Gandhi's temperance movement influenced much on women because they did know they suffered the humiliation of their fathers, brothers, and sons getting drunk.

Constitutional Provision

Article 47 of the Indian Constitution says that It is the duty of the State to raise the level of nutrition and the standard of living and to improve public health. The State shall regard the raising of the level of nutrition and the standard of living of its people and the improvement of public health as among its primary duties and, in particular, the State shall endeavour to bring about prohibition of the consumption except for medicinal purposes of intoxicating drinks and of drugs which are injurious to health. After independence first Prime Minister of India Jawaharlal Nehru even though a follower of Gandhi, refused to accept the complete prohibition of alcohol in the Country instead kept it on the State List most of the states did not want to put prohibition because did not want to face revenue cut short from that source. Gandhi's own son Harilal who was an opponent of Gandhi and became an alcoholic became a critique of the Gandhian concept. In Gujarat, Nagaland, Mizoram, and other states like Bihar banned alcohol in the state. Failing the Father of the Nation to keep India alcohol-free, Gandhi approached to keep India liquor-free. Bihar Chief Minister Nitish Kumar has banned alcohol in the state. Despite the loss of crores of rupees to the government, Gandhiji's commitment has been implemented. Instances of murder, gang robberies, traffic accidents, and domestic violence have gone down by 20 to 30 percent. (Indian Express, 4 August 2019) This is ironic that Liquor shops are running rampant in the streets named after Gandhi. Mahatma Gandhi Road in Bangalore. In Movies, Alcohol is used as evil used by a bad person's character or role. But nowadays using alcohol by heroes of movies and serials is quite common influencing youngsters that it is not at all a bad thing, therefore adults are using more alcohol **Findings:** India is the ninth-largest consumer of all alcohol in the world. The majority of the road accident deaths were attributed to drunk driving. Alcohol consumption is responsible for about 70% of violence in general and nearly 80% of domestic violence. Alcohol consumption increased by 55% in the last 20 years and is responsible for about 80% of domestic violence. The latest causal relationships have been established between harmful drinking and the incidence of infectious diseases such as tuberculosis as well as the course of HIV/AIDS. Alcohol consumption can do harm permanently to mental and physical disorders. **Recommendations:** Alcohol-related advertisements should be banned in all entertainment sectors like Cinema, TV, the Internet, and social media, etc., nowadays the popular media favours lurid descriptions of alcohol-related violence and heroic accounts of sporadic, it should be avoided.

Our decisions need to be strong enough to gain a healthy and wealthy life. Youth, Indians should follow the footprints of Mahatma Gandhi is essential to solve the issue **Conclusion:** According to youth who are consuming alcohol Gandhian philosophy is outdated but it is not outdated, for so many problems we have the solution in Gandhian Thoughts. Protests by women who are victims of their husbands, brothers, and sons of drinking have not attracted much to the public eye. Major Political Parties should accept to implement a ban on alcohol in their manifesto. God has given us this body, how can we pollute with alcohol? Who gave that authority to human beings? How we got the good body from the god, in the same manner, it is our duty to give back to the god in a correct manner till death.

References

1. Mahatma Gandhi, the Storey of My Experiments with Truth-An Autobiography

2. World Health Organization Report-2020

3. David M. Fahey and Padma Manian, The historian, FALL 2005, Vol.67, No. 3, PP 489-506, Taylor & Frncis Ltd

4. David Sherman.M, "The 10 most common causes of alcoholism", sanalake, 2020, https://sanalake.com/blog/the-10-most-commmon-causes-of-alcoholism/.

5. Mahima Nayar, In the shadow of alcohol: Women's experience of Bangalore Indian anthropologist, Vol. 39, No. 1/2 (Jan-Dec 2009) PP 49-64

6. Saxena.S, "Country profile on alcohol in India", Alcohol and public health in 8 developing countries, vol. 8, 1999, PP. 37-60

7. https://www.ambrosiaindia.com/2021/02/4494/

8. Turning Point, "5 effects of Alcoholism on family", https://turningpointtreatment.org/blog/5-effects-alcoholism-on-family/

9. An Empirical Study on the Role of Communication Skills Significant to the Entrepreneurs in Achieving Success.

Vijaya.B

Assistant Professor, Department of Commerce, GFGC, KR Puram, Bangalore, Karnataka, **India**

Abstract: Unemployment problem of a populous and developing economy like India can find its solution in entrepreneurship. However, an entrepreneur can be successful only when he/she is excellent in communication. Being one of the major elements in soft skills, communication is an integral part of the business not only at the time of establishing but also in running it profitably. Entrepreneurs need communication skills since they help them in sharing and communicating their ideas more clearly, as well as operate more effectively with their workers, team members, clients and investors. Good communication skills will also benefit an entrepreneur during project explanations, presentations, persuasion, training, leading, motivating and many other situations whenever a person interacts with others face to face and even on virtual platform. They have to recognize the importance of communication skills in business and equip themselves with some helpful hints for making their daily interactions, both internal and external, more productive. Unfortunately, the relevance of effective communication skills is ignored when compared with the other skills considered crucial to run the business such as organizational skills, leadership skills, business management, time management, creative thinking and problem-solving etc. This made me to undertake a study about the role played by communication skills in achieving success in an enterprise. A google form questionnaire was served to the entrepreneurs operating in Udupi district to know about their opinions regarding the significance of communication skills in business. The paper concludes by recommending the inculcation of effective communication skills by the entrepreneurs to enable them to run their enterprises successfully in a conducive environment for start-ups in India.

Keywords: Communication, Soft Skills, Entrepreneurs, Start-ups

Introduction

The distinct feature that distinguishes human beings from other living beings is their communication skills, especially, their speaking ability. People use communication to share and express their feelings, opinions, thoughts, needs etc. to others. The need for the same in business is really vast. An enterprise cannot be started and run without using communication. It is very crucial at all stages of enterprise right from mobilizing funds, assembling resources, conceptualizing the idea, developing it, presenting the project to the clients, closing sales etc., The managerial activities such as planning, organizing, staffing, directing, coordinating, controlling are impossible without communication. In this context, a person who wants to setup a business enterprise has to recognize the significance of communication in an enterprise and should make efforts to get awareness about the various communication skills as well as master the same. An entrepreneur must also know the various barriers such as physical, semantic, psychological, cultural, interpersonal and gender barriers that block the free movement of communication in an enterprise and overcome the same for securing effective communication.An investment of time and money in communication surely assists in realizing the best results of an enterprise.

Research Methodology: Statement of the Problem: Human beings need communication not only in their personal life but also in their professional life. Though communication is the most important life skill, yet many entrepreneurs knew only few ways of communicating and don't employ all the communication skills which are indispensable for realizing the goals of the enterprise. This fact made me to undertake the study of the role played by communication skillswhich are significant for achieving success.

Objectives of the Study:

1. To know the level of awareness about communication skills by the entrepreneurs.

2. To evaluate their knowledge regarding barriers to communication.

3. To help them recognize the significance of communication skills in the business.

4. To suggest the ways for improving their communication skills.

Research Design: Primary Data: The primary data was collected with the help of a suitable google form questionnaire whose link was sent to 30 entrepreneurs operating in Udupi district and each respondent was personally assisted over mobile phone call while filling the questionnaire and also the purpose of the study was explained. **Secondary Data:** The already existing secondary data was collected from the library along with the previous research work in the same area and of course with the help of internet sources. **Sample Size:** The number of respondents selected for the study were 30 entrepreneurs operating in various areas of Udupi district. **Method of Sampling:** It would be time consuming to prepare a sample frame of all entrepreneurs. Hence, Udupi district is selected for the present study due to close acquaintance and proximity to the research. **Plan of Analysis:** The raw data collected from respondents was tabulated and analyzed using simple mathematical and statistical techniques like addition, mean, frequency and percentage calculation.Based on the analysis of data collected from the respondents, the findings of the study were interpreted and suggestions were given. **Limitations of the Study:** The limitations of the study arise mainly because of time and resource constraints. Further, the findings of the study may change over a period of time as experience may polish the communication skills of the entrepreneurs over a period of time.

Review of Literature: An entrepreneur needs various skills like organizational skills, leadership skills, business management, time management, creative thinking and problem-solving etc. to run the enterprise successfully. Among these, communication skills are the most crucial. The English term communication has its roots in a Latin word "communis" which means common. "Communication is an exchange of facts, ideas, opinions or emotionsby two or more persons" –(W.H Newman and C.E Summer, The Process of Management). Communication is successful only when the receiver understands in the same sense and spirit in which the communicator intends to convey. For achieving this, an entrepreneur is expected to possess the following Communication skills. **Speaking:** Being the most fundamental form of communication, speaking is required for an entrepreneur for issuing orders, instructions, suggestions, warnings, advices, counselling, appreciations, motivation as well as for conducting meetings, training, workshops, giving speeches, information, sending voicemail, telephone/mobile conversation, etc., **Listening:** This is the most neglected skill in communication. If due consideration is given, it can be an effective weapon in the armor of an entrepreneur. It is very crucial for listening to the complaints, appeals, representations, at the time of counselling and mediation. In order to help customers, investors and employees, entrepreneur must pay full attention while listening to them and understand what is their expectation. Attentive, active and empathetic listening can help in picking up on current or potential problems early on, mitigating risks, building relationships, leading a team, and in resolving conflicts. **Reading:** It includes the capacity to perceive and interpret various documents, reports, business related information including diagrams, directories, correspondence, manuals, records, charts, graphs, tables, and specifications in a business setting. It helps an entrepreneur to get an overview, to locate specific information, to identify the central idea of the theme and to develop a detailed and critical understanding. **Writing:** Written communication allows businesses to communicate developments, expectations and legalities to employees and contracts in the outside world. Though business still keeps the hard copies of written communication, the use of email and online communication is increasing.Written communication, generally used with clients or other businesses, includeemail, internet websites, memos, bulletins, proposals, telegrams, faxes, postcards, contracts, articles, text messages, blog posts, social media posts,business letters, reports and job descriptions. **Body language:** Body language/kinesics is a type of communication in which physical behaviors such as facial expressions, body posture, voice, gestures, eye movement, touch and the use of space are used to express or convey the information. Attentiveness can be expressed by turning the head towards the speaker, making eye contact, leaning forward, nodding the head and avoiding mobile during listening situations. Handshake and a smile can do a lot in creating the best impression. **Presentation:** Presentation is a type of representation of an entrepreneur's ideas, proposals, data or other information through pictures to the clients, investors, journalists, employees or other groups. Entrepreneurs may present accounts of their enterprises to the investors, business services to the interested customers or give speech about entrepreneurship to the students in an educational institution. A picture speaks thousand words and relates itself with the listeners orally as well as visually and always yields fruitful outcome. Effective presentation requires clarity, conciseness, confidence and poise and surely assists in the growth of the business. **Mediation:**Mediation is a voluntary collaborative process where individuals who have a conflict with one another identify issues, develop options, consider alternatives and develop a consensual agreement. An entrepreneur facilitates open communication to resolve differences in a non-adversarial and confidential manner. The conflict may be between two employees or between two competing vendors or between any two or more persons with different opinions. The objective of mediation is to let the parties strike a settlement in case of Payment dispute, Contract dispute or Supply dispute. **Cross – platforming:** Information can be passed through so many channels such as telephone, mobile, email, instant messages, video chat, etc. The challenging task in front of entrepreneur is to choose which medium is appropriate for the message to be sent. Proper choice of medium and sending the message through it results in successful transmission of the message. **Barriers to Communication:** Communication barrier is a block which prevents the transmission of the message from reaching the recipient. The message may be distorted, misinterpreted, poorly translated, may not reach the target audience or only a part of the message may be conveyed or important content may be missed out because of Communication barriers such as Physical,Semantic, Psychological, Interpersonal, Cultural and Genderbarriers.

Significance of Communication Skills to the Entrepreneurs:

Business communication skills are very crucial in every aspect of business. A lot of HR complaints can be resolved, or avoided altogether, through good communication.

In addition to office relations, good communication skills also allow:

1. Promotion of innovation and creativity
2. Improved marketing and promotion for the business
3. Inspiring team members, colleagues, and employees towards goals or business aspirations
4. Improving customer experience and retaining them
5. Reduced stress and confusion
6. Facilitates decision making.
7. Enhanced productivity
8. Improves overall business development
9. Strengthens the bond with clients and stakeholders

10. Build positive relationship with the employees.

Good communication skills can enhance office and work relationships, speed up processes, and smooth disagreements. By contrast, poor communication skills can result in misunderstandings, lost time, arguments, and even lost money or business because of missed deadlines or confusing information.

Data Analysis and Interpretation: Table 1.1 Awareness about Communication Skills

Communication Skills	Yes	No
Speaking	30	0
Listening	25	5
Reading	25	5
Writing	24	6
Body language	22	8
Presentation	18	12
Mediation	10	20
Persuasion	10	20

All entrepreneurs are aware about Speaking skill. Most of the entrepreneurs are not aware about Presentation, Mediating and Persuasion skills.

Table 1.2 Awareness about Barriers to Communication

Barriers	Yes	No
Physical	18	12
Semantic	10	20
Psychological,	16	14
Interpersonal	12	18
Cultural	13	17
Gender	14	16

Semantic barriers are not known by two third of the entrepreneurs followed by interpersonal barriers, cultural barriers and gender barriers. However, physical barriers are well known by them.

Table 1.3

Significance of Communication Skills to Business

Communication Skills	Most Significant	Significant	Least Significant	Neutral
Speaking	28	-	2	-
Listening	20	8	2	-
Reading	10	18	2	-
Writing	14	12	4	-
Body language	6	14	6	4
Presentation	16	-	6	8
Mediation	6	10	6	8
Persuasion	6	10	4	10

Speaking is the most significant communication skill followed by listening, presentation and writing. On contrast, body language, mediation and persuasion are not considered most significant.

Table 1.4 Ways of improving Communication Skills

Ways	Strongly Agree	Agree	Neutral	Disagree	Strongly Disagree
Regular Reading	14	**12**	2	2	-
Active Listening	**18**	10	-	2	-
Practice before Speaking	14	10	4	2	-
Join a Course	10	10	8	2	-
Attend training	18	8	-	4	-

program					
Listen to podcasts	10	8	6	6	-
Use mobile apps for self-study	4	16	6	2	-
Sticking to the main points	12	12	2	4	-
Articulating good body language	12	10	4	4	-
Showing respect	18	10	-	2	-
Using proper medium	10	20	-	-	-
Transparency	14	14	2	-	-
Hire an expert	6	8	10	6	-

Most of the entrepreneurs strongly agree that active listening and attending training programs, showing respect, regular reading, practice before speaking and transparency are the most significant ways for improving communication skills.

Findings:

1. Most of the respondents are men.

2. Half of the respondents belong to the age category of 25 years to 35 years.

3. All respondents agreed that communication is essential to the business.

4. Body language, presentation, mediation and persuasion skills are not known by many respondents.

5. One third of the entrepreneurs are not aware of the barriers to communication.

6. Speaking is the most significant communication skill. Listening, writing and reading occupy the next positions. However, the remaining communication skills are not considered significant.

7. Most of the entrepreneurs said that it is essential to master the communication skills.

8. Regular reading, active listening and attending a training programs are the ways in which communication skills can be improved.

9. All respondents agreed that communication skills can be taught during the student days.

Conclusion: Entrepreneurs consider communication taken for granted without giving due relevance to it. Most of them do not know how to frame the message, the language and the words they have to choose, the medium through which it has to be sent, the barriers that prevent the effective transmission of the message as well as the significance of communication skills in the business. Nobody can be an effective communicator by birth itself. However, the awareness and continuous practice can make one such.

Suggestions:

1. Entrepreneursshould get an awareness about Body language, presentation, mediation and persuasion skills.

2. The knowledge about communication barriers and the ways of overcoming is must for entrepreneurs.

3. The communication skills can be mastered by active listening, attending training programs, regular reading, practice before speaking, articulating good body language etc.

4. Proper use of National Education Policy to inculcate communication skills during their student hood.

References:

1. Khanka, S.S. (2012), 'Entrepreneurial Development', S. Chand & Company Ltd., New Delhi.

2.Varinder Kumar and Bodhi Raj (2001), Business Communication, Kalyani Publishers, New Delhi.

10. Our Experience with Some Rare Parotid Masses.

Dr Keshav Gupta[1] and Dr Ashu Bhati[2]

[1]Assistant Professor, Department of ENT and Head & Neck Surgery, GS Medical College & Hospital Hapur India

[2]Postgraduate Resident, Department of ENT and Head & Neck Surgery, GS Medical College & Hospital Hapur India

Abstract: The parotid gland is a major salivary gland in the human body located in both the cheeks. It accounts for approximately 3-6% of all head and neck masses. Three fourths of all parotid masses are benign. However, they are very diverse in nature and in intimacy with important anatomical structures. An accurate diagnosis and surgical management of some rarely presenting parotid masses is further challenging to the medical team because of limited literature and experience with these masses. Here we present different approaches towards rarely presenting parotid masses. A giant pleomorphic adenoma with 14 cm diameter managed with total superficial Parotidectomy, a difficult to diagnose parotid mass at angle of mandible which was confirmed to be Schwannoma post excision and a large Paediatric lymphangioma managed with Sclerotherapy are discussed here. All were managed without mortality, with minimal morbidity and no recurrence noted till date. Our experience will be of great use to physicians and surgeons while dealing with rare benign parotid masses.

Keywords: Parotid Mass Swelling Management Diagnosis Surgery

Introduction and Review of Literature

Human body houses many salivary glands which release their secretions through the ducts opening into the oral cavity. These secretions are composed of saliva, enzymes, minerals, water and immunoglobulins and hence responsible for the mechanical, defensive, digestive, immunological and pH regulatory functions of the oral cavity. There are larger paired major salivary glands and smaller numerous minor salivary glands. Parotid gland is the largest of all salivary glands which houses both the cheeks. Its secretions are watery and reaches the oral cavity through its duct named as Stenson's duct. The parotid gland is inverted pyramidal in shape with its apex towards the angle of mandible and its base at the zygomatic arch anterior to the external auditory canal. Anteriorly it is in relation with buccal pad of fat and posteriorly it curves around the posterior border of mandible (Malik, 2016). The gland is covered by a tough, non-yielding false fascia which is formed by condensation of deep fascia of neck and it splits to surround the gland. The gland is composed of a superficial lobe and a deep lobe. Important anatomical structures like Facial Nerve, retromandibular plexus of veins, Facial Vein, external carotid artery and its branches passes through the substance of the gland. Facial nerve enters the gland as soon as it leaves the stylomastoid foramina either in form of a single trunk or its branches and gives off its terminal branches within the substance of the gland. The Stenson's duct emerges at the anterior part of the gland, passes horizontally through masseter, pierces buccinator and opens in oral cavity through papillae on both sides opposite the upper second molar teeth (Chaurasia, 2023). Parotid masses are quite common and accounts for approximately 3-6% of all head and neck masses. Majority (almost three-fourths) of all parotid masses are benign masses and rest are malignant (Tartaglione, 2015). Benign parotid masses also consist of a large variety of masses and have been classified on various basis. One of the largest studies on benign parotid masses were conducted by Bradley and Mc Gurk (2013) (1065 cases) and Everson and Cawson (1985) (2410 cases). The incidence of benign parotid masses was reported to be 5.3-6.2 per lakh population. No age group is spared. Sixth decade of the life is most likely to face a benign parotid mass. Females are more likely to present with this condition (Homer et al, 2019). Diagnosis as well as treatment needs good clinical as well as technical expertise. Pleomorphic adenoma (PA) is one of the commonest parotid masses. PA has a definitive anatomical relation between Facial Nerve (FN) and its branches and hence its surgical management may be challenging. A large PA is more challenging as it may be in close relation with many or all branches of FN (Xie et al, 2015). Schwannoma are rare benign parotid masses. There are limited studies available in literature about Parotid Schwannoma (PS). Large Bradley- Mc Gurk and Everson-Cawson studies did not report a single case of PS. Parotid Lymphangioma (PL) are another rare benign parotid mass. Management of paediatric parotid masses is always challenging as they require dedicated and specialised medical care right from presentation till the last step in management. Surgery may be quiet challenging in this age group. There are various factors which needs to be considered while managing a case of rare benign parotid mass. Prompt diagnosis in a systematic and cost-effective manner in a single sitting may be challenging in such cases. After a diagnosis is reached, the next step is to decide the best treatment which may be medical or surgical or both. Surgical cases require choosing the best approach, meticulous planning, good surgical exposure, good surgical skills, perfect completion and team work. We have addressed all these factors for the management of benign parotid mass in detail in the present paper.

Aim: To study the management of rare benign parotid masses.

Objectives: To study the best diagnostic modalities in prompt, systematic and cost-effective diagnosis of rare benign parotid masses To study the treatment options available in management of benign parotid masses. To decide the best surgical approach in surgical management of rare benign parotid masses

Material and Methods

We conducted an observational retrospective study at our Tertiary care Hospital to study the management of rare benign parotid masses in indoor patients in Department of Otorhinolaryngology and Head and Neck Surgery during last two years. Informed and written consent for publication for using the clinical data in good faith were taken. Patient's right to confidentiality has been maintained and patient's identity is revealed nowhere. The clinical knowledge hence gathered has been reviewed with knowledge from the literature and present protocols to conclude the right management approach for management of rare benign parotid masses. Inclusion criteria- Patients of all sexes and all age groups who were admitted in Department of Otorhinolaryngology and Head and Neck Surgery at GS Medical College & Hospital Hapur India from April 2021- March 2023 diagnosed with rare benign parotid masses were included in the study Exclusion criteria- Patients with common parotid swellings, infectious conditions and malignancy were excluded from the study.

Observations: The first case was that of a 34 years male skilled labourer by occupation who presented with a right parotid mass of 12 years duration (Figure 1.1). The swelling was of gradual onset and slowly increasing in size. It was a single painless swelling in the region of right cheek. There was no history of sudden increase in size, any episode of decrease in size, bursting, discharge, change in overlying skin, abnormal sensations. There was no history of trauma, radiation exposure, exposure of long duration to heat, heavy metals, dyes and toxins. However, there was long duration exposure to bright sunlight almost 8 hours a day for last 20 years. There was no previous history of occurrence and/or resolution of a noticeable long duration swelling in same or any other region of the body. There was no history of fever and weight loss. There was no history of systemic illnesses. Patient is a chronic smoker and smokes 2-5 bidis a day for last 10 years. The mass was measured to be an ovoid mass with maximum 14 cm diameter. It had a smooth surface, no edges, no pulsations and no impulses. Temperature of overlying skin was normal. There was no tenderness, well defined regular margins, no fixity to skin or other structures. It was non-compressible and non-reducible. The mass was dull on percussion with no auscultatory sounds. There were no palpable lymph nodes. Bilateral Facial Nerve was intact. Figure 1.1: A large parotid mass of 12 years duration in a 34 years old male patient. An ultrasound guided FNAC followed by CECT Neck (Figure 1.2) were done in a single sitting. All investigations suggested PA and demanded histopathological

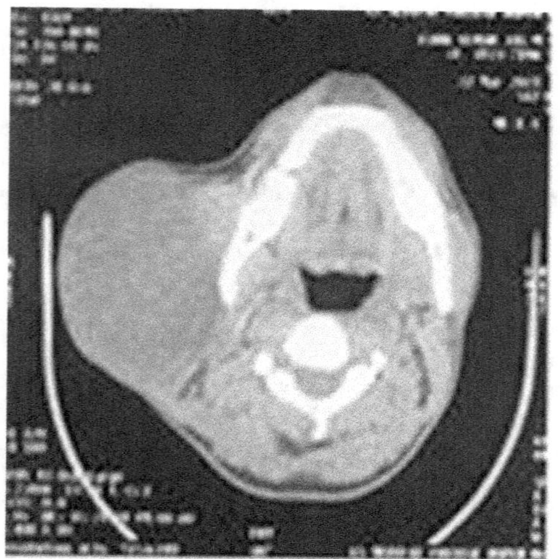

confirmation.

Figure 1.2: Contrast enhanced computerised tomography (CECT) shows a large mass in superficial lobe of parotid in close contact with important neurovascular structures, particularly the branches of the Facial Nerve (FN). The second case was that of an 18 years male student who presented with a very slowly progressive left infra-auricular mass over last 5 years (Figure 1.3). Rest of the mass was freely mobile but there was slight indentation on skin at a single point at its periphery. There were no overlying or surrounding fistula or sinus. However, there was no history of tuberculosis in patient and close relatives but it still could not be ruled out at presentation. A FNAC done 3 months back suggested a benign inflammatory lesion with epithelioid cells? Tubercular. However, diagnosis could not be confirmed. We did a USG and core biopsy at our centre. USG showed a hypoechoic lesion in superficial lobe of parotid gland with a solid cellular structure without a cavity. Histopathology was recommended to find its exact nature. Histopathology showed cells which were well stained with eosin only without any nuclear material. No signs of malignancy, necrosis and inflammation were seen. Excision was advised to find its exact nature. Figure 1.3: A small infra-auricular with a skin indentation at a single point in an 18 years old male patient for last 5 years The third case was that of a 2 years male child with a huge swelling of left cheek (Figure 1.4). It was present almost since birth and gradually increasing in size since then. It blenched on digital pressure and had no pulsations. USG with colour doppler, USG guided FNAC and CECT neck were all suggestive of PL. Figure 1.4: A large parotid swelling in a 2 years old male patient present and enlarging since birth.

Results : The first case was a large parotid PA. Contrast enhanced computerised tomography (CECT) showed that the mass was limited to superficial lobe of the parotid gland. It was in intimate relationship with four branches of the FN on its deep surface. It was excised in toto by total superficial parotidectomy (Figure 1.5-1.6). The intra-operative period was uneventful. There was transient FN dysfunction in immediate post operative period which was probably due to handling of the branches of the FN. The FN function completely recovered within 72 hours after surgery spontaneously. Histopathology confirmed the diagnosis of PA. There has been no recurrence till date. Total superficial parotidectomy after complete removal of superficial lobe of parotid gland. Facial Nerve (FN) and its branches are intact in the bed. The Pleomorphic Adenoma (PA) of maximum 14 cm diameter was excised intact and complete, and sent for histopathological evaluation. Pre-operative diagnosis was not clear in the second case. It was a small mass away from the major neurovascular bundles. Partial superficial parotidectomy was done in this case and the mass was completely excised with 5 mm strip of normal surrounding tissue and sent for histopathology. There were no intra-operative and post-operative complications. Histopathology revealed presence of compact hypercellular Antoni A and myxoid hypocellular Antoni B areas suggestive of PS. Partial superficial parotidectomy was done with 5 mm strip of normal tissue around the mass and sent for histopathology. The third case was a paediatric case with PL. USG doppler was done to rule out arteriovenous malformation, presence of blood flow within the mass and significant direct vascular contribution to it by the surrounding blood vessels. CECT Parotid was done to note topographic and angiographic details. The child was taken to the paediatric operation theatre where under sedation, anaesthetic monitoring and after sensitivity testing under CT guidance, a 27 Gauze needle was passed into the mass perpendicular to the skin in pre auricular region, thick red blood was aspirated from the collection of immature vessels and 1 mL of sclerosing agents were injected. The child was kept under monitoring for 24 hours and discharged the next day with a plan for repeat procedure after a month for the residual mass. There was no residual mass when the patient came for follow up after 7 days (Figure 1.8). The child is under regular follow up for last 3 months and no recurrence is seen during this period. Also, no repeat procedure was done during this period. The mass completely disappeared on first follow up 7 days after the first session of sclerotherapy.

Discussion

Aetiology of benign salivary gland tumours is not well understood. There have been concerns regarding petrochemicals, dyes (Young, 2022), radiation (Saku et al, 1997), mobile phones (Soderqvist et al, 2012) and Human Papilloma Virus (HPV) (Skalova et al, 2013) but none has a proven association with benign parotid tumours. Benign parotid tumours are slow going and usually devoid of other symptoms other than the presence of a mass. Presence of pain, abnormal sensations, increase in growth rate, weakness, skin involvement, fixity or irregularity should alert the physician for possibility of malignant transformation. It is advisable to do an ultrasonography (USG) and USG guided FNAC at presentation in a single sitting. It saves time, money and provides a more accurate diagnosis. USG is very useful as it increases accuracy of FNAC, fulfils most of the requirements of imaging (Gritzmann et al, 2022), distinguishes a lump from a node and from diffuse enlargement of the gland. A CECT/ Magnetic Resonance Imaging (MRI) should be done in cases of tumours larger than 3 cm, involvement of deep lobe and parapharyngeal space and suspicion of malignancy. FNAC has an accuracy of 80-90% (Al-Khafaji et al, 1998) and a core biopsy (Douville et al, 2013) increases sensitivity to 100% and specificity to 92% (Romano et al, 2017). In our case of PS, a single point of indentation on skin may be attributed to previous FNAC which was done 3 months back. Common differentials of parotid masses are PA, Warthin's tumour, Oncocytoma, Monomorphic masses, basal cell carcinoma (BCC), cystadenoma and myoepithelioma (Naggar et al, 2017). Surgery is indicated in most of the parotid masses for definitive histology, presence of continued growth in persistent mass and probability of malignant transformation. Enucleation is not performed now because of high recurrence rate (Bradley et al, 2012). Either total superficial parotidectomy or partial superficial parotidectomy with a cuff of normal tissue should be done (Xie et al,

2015). A parotid tumour will always be found near the FN and its branch (McGurk et al, 1996) and hence carries high risk of iatrogenic FN injury. Overall recurrence rate of PA is 2% and usually years after the surgery (Valstar et al, 2017).

Conclusion

All parotid masses should be managed in a systematic and a comprehensive manner. History taking and clinical examination are the cornerstone of the management. USG is the most valuable imaging modality. USG with FNAC should be done in a single sitting. However, CECT and MRI may be required in select cases. Core biopsy may add on to histopathological evaluation. Total and partial superficial parotidectomy are the preferred surgical techniques for the masses of superficial lobe. Surgery has to be done very carefully as in most of the cases FN or its branch will be in close relation. Sclerotherapy is a promising novel technique for angiogenic masses. Long term follow up is to be done as recurrence may be seen decades after surgery.

Refrence

1. Al-Khafaji BM, Nestok BR, Katz RL (1998). Fine needle aspiration of 154 parotid masses with histological correlation: ten year experience at the University of Texas M.D. Anderson Cancer Center. Cancer, 84(3), 153-159.

2. Bradley PJ, & Mc Gurk M (2013). Incidence of salivary gland neoplasms in a defined UK population. Brazilian Journal of Oral and Maxillofacial Surgery, 51(5), 399-403.

3. Bradley PT, Paleri V, Homer JJ (2012). Consensus statement by otolaryngologists on the diagnosis and management of benign parotid gland disease. Clinical Otolaryngology, 37(4), 300-304.

4. Chaurasia BD (2023). Human Anatomy. New Delhi: CBS Publishers.

5. Douville NJ, Bradford CR (2013). Comparison of ultrasound-guided core biopsy versus fine-needle aspiration biopsy in the evaluation of salivary gland tumours. Laryngoscope, 127(11), 2522-2527.

6. Everson JW, & Cawson RA (1985). Salivary gland tumours. A review of 2410 cases with particular reference to histological types, site, age and sex distribution. Journal of Pathology, 146(1), 51-58.

7. Gritzmann N, Hollerweger A, Macheiner P, Rettenbacher T (2022). Sonography of soft tissue masses of neck. Journal of Clinical Ultrasound, 30(6), 356-373.

8. Hommer J, & Robson A (2019). Scott Brown's Otorhinolaryngology Head & Neck Surgery: Benign salivary gland tumours. New York. IL: CRC Press.

9. Malik NA (2016). Textbook of Oral and Maxillofacial Surgery. New Delhi: Jaypee Publishers.

10. Naggar AK, John KC, Grandis JR, Takata T, Slootweg PJ (2017). WHO classification of tumours of salivary glands, 11(4), 160-184.

11. Romano EB, Wagner JM, Alleman AM (2017). Fine-needle aspiration with selective use of core needle biopsy of major salivary gland tumours. Laryngoscope, 127(11), 2522-2527.

12. Saku T, Hayashi Y, Takahara O, et al (1997). Salivary gland tumours among atomic bomb survivors, 1950-1987. Cancer, 79(8), 1465-1475.

13. Skolova A, Kaspirkova J, Andrie P, et al (2013). Human papillomaviruses are not involved in the etiopathogenesis of salivary gland tumours. Cesk Patol, 49(2), 72-75.

14. Soderquist F, Carlberg M, Hardell L (2012). Use of wireless phones and the risk of salivary gland tumours: A case control study. European Journal of Cancer Preview. 21(6), 576-579.

15. Tartagoline T, Botto A, Sciandra M, Gaudino S, Danieli L, Parrilla C, Paludetti G and Colosmino C (2015). Differential diagnosis of parotid gland tumours: which magnetic resonance findings should be taken into account? Acta Otorhinolaryngology Italy, 35(5), 314-320.

16. Valstar MH, de Ridder M, van Den Broek EC et al (2017). Salivary gland pleomorphic adenoma in the Netherlands: a nationwise observational study of primary tumour incidence, malignant transformation, recurrence and risk factors for recurrence. Oral Oncology, 66(1), 93-99.

17. Xie S, Wang K, Xu H, et al (2015). PRISMA- extracapsular dissection versus superficial parotidectomy in treatment of benign parotid tumours: evidence from 3194 patients. Medicine, 94(34), e1237.

18. Young A, Okuyemi OT (2022). Benign Salivary Gland Tumours. In: StatPearls[Internet]. Treasure Island. https://www.ncbi.nlm.nih.gov/books/NBK564295/

11. Enhancing the Learning of Induction Machines through MATLAB.

Abdulhamid Musa

Electrical and Electronic Engineering Department, Petroleum Training Institute, Effurun, Delta State, **Nigeria**.

Abstract: By utilizing MATLAB as a research instrument, this investigation aims to get a higher level of comprehension and knowledge acquisition about induction machines. In addition, this research offers a strategy that may be used to deal successfully with the complexities associated with the performance of these machines. In the field of modelling and control approaches associated with induction machines, significant development has been made throughout the course of time. In addition to static and dynamic modelling of three-phase induction machines, the achievements include sensor less indirect field-oriented control merged with fuzzy logic. MATLAB/Simulink has been shown to be an effective instrument for carrying out these approaches by many studies, which have produced evidence of this effectiveness on multiple occasions. Therefore, it is necessary for future research in the field of induction machine modelling and control to place a greater emphasis on developing intricate control algorithms, investigating novel intelligent methodologies, and improving existing models to accommodate potential flaws and high-performance circumstances. In addition, conducting additional research into the testing and characterization of induction devices is essential.

Keywords: Induction Machines, MATLAB/Simulink, Simulation, Optimization, Fault diagnosis

Introduction

Induction machines (IMc) are commonly used in various industries because of their simplicity, robustness, and reliability. These electrical machines operate based on the principle of electromagnetic induction. They are used in applications such as pumps, fans, compressors, and conveyors. In simple terms, IMc converts electrical energy into mechanical energy. They are mostly used in several industries due to their efficiency and cost-effectiveness. Two main types of IMc based on their rotor constructions are the squirrel cage and the wound rotor. The squirrel cage IMc is the most used type because it is simple and requires low maintenance. On the other hand, the wound rotor IMc is used in applications that require variable speed control and a higher starting torque [1]. Both IMc can be modelled and controlled using MATLAB, a powerful tool for the simulation and analysis of electrical systems. MATLAB is a widely used software tool for modelling, simulation, and analyzing electrical systems, including IMc. It provides a user-friendly interface for designing and testing IMc control systems [2]. MATLAB is software that simulates different IMc tests, including DC, no-load, and blocked-rotor tests. These tests can help identify the equivalent circuit parameters of the machines [3, 4]. Additionally, MATLAB can be used to teach and simulate adjustable speed drives of IMc, making it an essential tool for electrical engineering students and professionals [1, 2] [5, 6]. This study explores the use of MATLAB as an educational tool for teaching IMc, focusing on its benefits in understanding induction machine theory, supporting effective design and simulation, and its effectiveness in studying these machines. It also discusses the significance of IMc, MATLAB's potential in enhancing learning, and various simulation instances demonstrating its efficacy in studying these machines.

Literature review

This study investigates the use of reference frames in electrical machine evaluation. It presents a simple model for a three-phase induction motor (IMt) using dq0 axis transformations, demonstrating that reference frame theory can accurately demonstrate the steady-state operation of IMc [7]. However. Another study by
[8] developed a mathematical model of an operator workstation, illustrating user interface elements, formalizing on-screen controls in human-machine interaction systems, and developing structural-functional models for parameter measurement and synchronous generator system monitoring. The utilization of electronic devices and microprocessors in electrical motors facilitates the achievement of enhanced braking performance with asynchronous motors. The study introduces a rapid deceleration method for achieving a full halt of IMt. This methodology has been verified through simulation experiments

conducted across various operational scenarios [9]. This study explores three-phase induction motor dynamics and transient conditions in low, medium, and high-power motors using MATLAB/Simulink and a mathematical model. It measures voltage, current, flux, speed, and torque and uses a stationary reference frame q-d axis model for testing 5, 50, and 500-hp IMt in various scenarios. [10]. Field-oriented induction motor regulation requires rotor flux measurement. Induction motor rotor flux measurement is difficult. Neural networks estimate rotor flux in this study. Induction motor model equations simulate rotor flux compared to the neural network's output. [11]. The study introduces a new induction motor start-up method, eliminating torque pulsations and damaging mechanical system components. It uses MATLAB/Simulink models, SKT 24/12E thyristors for AC voltage controllers, and 89C51 microcontrollers for gate control and simulation verification [12]. The study introduces a step-by-step Simulink-based induction machine model creation method, which can be implemented using open-loop constant V/Hz control and indirect vector control [13]. Current or voltage models estimate flux for induction motor direct vector control. The current model operates at low speeds, while the voltage model flux estimator works at high speeds. Wide-speed hybrid estimators are recommended. Rotor flux-oriented control was studied. Simulations show the hybrid estimator improves flux estimates [14]. This paper presents "Enhancing the learning of IMc through MATLAB." The aim of this paper is to present a proposed approach for reviewing learning methods of IMc using MATLAB. It explores

IMc' intricacies and enhanced methods of learning IMc through MATLAB. Besides, it aims to provide a comprehensive and insightful analysis. The objective of this research is to demystify the complexities surrounding IMc. By leveraging the power of MATLAB, the intricacies of this subject matter can be researched, and a deeper understanding of the underlying principles can be gained.

Methodology

This section provides an overview of the techniques employed to clarify the principles behind IMc using MATLAB. To facilitate this review, the study is broadly categorized and presented in Figure 1.1. This figure visualizes the intended research design, providing important insights into the methodology employed and the variables under consideration. The figure illustrates several essential components for assessing the operational efficiency of IMc.

Figure 1.1:   Various   methods   used   for   the   study.

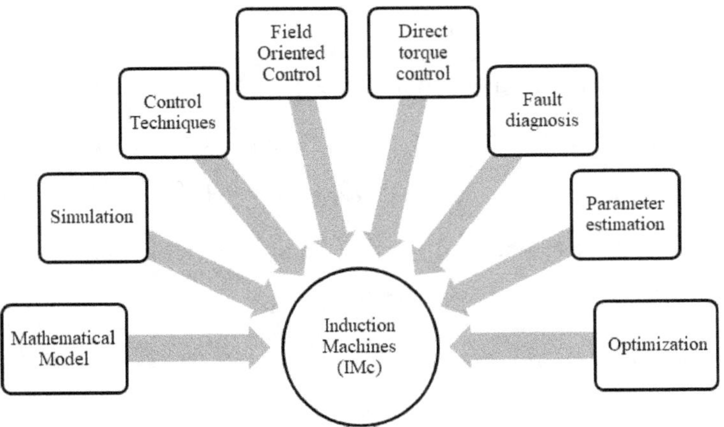

Results and Discussions

The proposed review's arrangement in Figure 1.1 is thoroughly examined, highlighting the study's methodology and the significance of each component. This comprehensive analysis demonstrates a valid and reliable research framework. The study of IMc modelling and control using MATLAB represents a rapidly evolving field with potential for future advancements. MATLAB/Simulink offers a powerful tool for modelling, simulating, and testing IMc, enhancing performance and efficiency across various applications.

Mathematical Model of IMc

Electromagnetic induction underpins the IMc mathematical model. A magnetic field and electric current generate a mechanical force. The machine's electromagnetic and mechanical systems are modelled using differential equations. It considers various parameters such as resistance, inductance, mutual inductance, moment of inertia, and friction [1]. Developing a mathematical model for IMc is a complex process that requires advanced mathematical techniques. Researchers have recently focused on creating accurate and efficient methods for modelling these machines. One such method is the field reconstruction method, which utilizes a small number of finite-element models to simulate the behaviour of the machine [15]. Another approach involves using Simulink/MATLAB models, which provide a powerful tool for simulating the behaviour of IMc and can be used for both teaching and research purposes [16]. Parameters of IMc are crucial for their behaviour and performance. MATLAB offers a powerful tool for determining these parameters through simulation models. An IMc can be determined using a MATLAB/Simulink implementation, transforming stator and rotor variables using dq0 axis transformations [17, 18]. Likewise, there are various methods for experimentally determining the parameters of IMc, such as using special tests through the MATLAB/Simulink program [19, 20]. Overall, the mathematical model of IMc and determining their parameters are essential for designing and analyzing these machines.

Simulation of IMc using MATLAB

MATLAB is a powerful tool for simulating and modelling electrical and IMc. It involves defining machine parameters like resistances, inductances, and mutual inductance and deriving machine equations like voltage, current, and torque to develop a simulation model [2]. MATLAB offers several built-in blocks for modelling IMc; the Induction Motor block implements a three-phase IMt. The simulation of IMc in MATLAB can be used for various applications, such as direct torque control, field-oriented control, and adjustable speed drives [21]. It can also be used to test and identify the machine Field's equivalent circuit parameters [19]. Additionally, MATLAB can be used for both static and dynamic modelling of IMc, including modelling of possible faults and high-performance applications. By simulating IMc in MATLAB, engineers can accurately design motors that meet protective standards for specific purposes [1]. To simulate a variable-speed induction motor transmission using MATLAB, the first step is to model the motor and its control system. This can be done using the built-in blocks in MATLAB/Simulink, such as the IMt and PI Controller blocks. The simulation can then be run to observe the performance of the motor and its control system under different operating conditions [4] [22]. Real-time implementation of IMt drives using MATLAB and digital signal processor boards is also possible and has been described in detail in the literature [5]. The versatility and flexibility of MATLAB make it an ideal tool for modelling and simulating IMc, allowing engineers to design and optimize these machines for various applications.

Induction Machine Control Techniques

In industrial settings, IMc are frequently utilized due to its robustness and reliability. To control these machines various control techniques have been developed over the years. An overview of induction machine control techniques involves developing algorithms for controlling IMc [21]. One of the primary objectives of induction machine control is to achieve high-performance operation while minimizing energy consumption. To achieve this, both static and dynamic modelling of the IMc is necessary [1]. MATLAB/Simulink is a powerful tool that can be used to develop and implement these control algorithms [2]. Scalar control techniques are one of the most used control techniques for IMc. The Induction Machine Scalar Control block controls the IMc using a scalar control structure, such as V/f or V/Hz. This block implements the control structure. [19]. The scalar control method maintains a constant voltage-to-frequency ratio to generate reference voltages from the reference frequency [20]. This method is simple to implement and requires minimal computational resources. The V/f control method can be used to regulate the speed of the rotor in an IMc drive powered by a matrix converter [21]. Another widely used control technique for IMc is vector control. Vector control techniques include Space Vector Modulator IM Direct Torque Control Induction Machine Field-Oriented Control and vector control schemes for voltage-fed inverter induction motor drives [21] [23]. Vector control techniques require more complex algorithms and computations than scalar control techniques but offer higher performance and efficiency. MATLAB/Simulink can be used to simulate vector-controlled IMc systems [24]. Implementing these advanced control schemes can help improve the performance and efficiency of IMc in various industrial applications [25].

Field-Oriented Control of IMc

Field Oriented Control (FOC) is a technique commonly used to manage IMc. The MATLAB Induction Machine Field-Oriented Controller block applies a FOC structure to IMc using the per-unit system. FOC enables precise control of the machine's torque and speed, making it an essential tool in modern industrial applications. The induction machine can use FOC to operate efficiently, reducing energy consumption and increasing productivity. To use FOC, it's crucial to understand the mathematical model of the system. This involves transforming the machine's three-phase variables into a two-axis reference frame, which allows for independent control of the machine's flux and torque. The FOC model also includes a closed-loop speed control based on indirect or feedforward vector control methods [27]. Moreover, an adaptive FOC scheme can be used to decouple the machine's flux and torque control [26]. Simulating and implementing the FOC system becomes easier with a good understanding of the mathematical model. MATLAB provides a powerful platform for simulating and implementing FOC. Simulink/MATLAB can be used to create a dynamic model of an IMt, which is useful for teaching electric machines and power systems courses. A video series demonstrates how to use Simscape Electrical to construct a model with a field-oriented induction motor controller. Furthermore, research has been conducted on developing high-performance vector control algorithms for IMc [26]. Another paper by [5] explained how to use Simulink blocks to teach and simulate an adjustable speed drive for an induction motor. The focus is on an indirect field-oriented control algorithm. Consequently, MATLAB provides a comprehensive platform for teaching and implementing FOC for IMc.

Direct Torque Control of IMc

Direct Torque Control (DTC) is a widely used method for controlling IMc. It is a high-performance control technique that can achieve fast torque and speed responses while reducing torque ripple. The IMc Direct Torque Control block in MATLAB implements the DTC structure for IMc. The figure shows the block diagram of the equivalent circuit [32]. This article discusses the mathematical model, simulation, and implementation of DTC for IMc. The control of IMc using DTC involves using stator flux and torque as control variables. This is done by changing the voltage vector applied to the machine's stator. A mathematical model of the IMt has been proposed in [27], which includes equations for the motor, stator, rotor flux, and torque. This model is used to design the DTC controller and simulate the performance of the IMc. Simulation and implementation of DTC for IMc can be done using MATLAB/Simulink. A study in [27] demonstrated a simulation of DTC for three-phase IMt using Field Programmable Gate Arrays (FPGA) technology. The simulation showed that the proposed DTC method effectively achieved fast torque and speed responses with reduced torque ripple. A virtual teaching platform for undergraduate students concerning an IMt drive technique is proposed in [28]. The platform allows students to simulate and implement DTC for IMc, providing a hands-on learning experience.

Induction Machine Fault Diagnosis

Fault diagnosis is an important aspect of maintaining the health of IMc. There are various fault diagnosis techniques that can be used to identify and diagnose faults in IMc. Machine learning techniques have been used to diagnose faults in electrical machines, such as IMc. This is done by analyzing and categorizing information at the edge [29]. MATLAB is a popular tool for fault diagnosis in IMc, generating data using a Simulink model and creating a multi-class classifier to detect defect combinations, enhancing accuracy and efficiency through machine learning techniques. Induction Machines are susceptible to various failure types, including broken rotor bars, bearing faults, stator faults, and supply voltage faults [30]. MATLAB's Diagnostic Feature Designer app can identify faults in AC IMc, specifically broken rotor faults, by analyzing vibration and electrical signals. The field has proposed a practical method for diagnosing faults in IMt using machine learning and experimental data [31]. The discrete wavelet transform has also been used for fault diagnosis of IMc [32]. Thus, MATLAB provides various tools and techniques for fault diagnosis of IMc. MATLAB has been used to model and simulate various fault conditions in IMc, including bearing failures and supply voltage variations [33] [34]. Accordingly, the behaviour of IMt under different operating conditions has been studied by implementing three tests (DC, no-load, and blocked rotor) using MATLAB/Simulink. Besides, a strategy based on machine learning has been proposed to identify faults in the power connections of IMc [35]. MATLAB provides a comprehensive platform for teaching and learning about induction machine fault diagnosis.

Induction Machine Parameter Estimation

Induction machines are crucial for industrial applications, and accurate parameter estimation is essential for efficient operation. Techniques involve specific tests and mathematical models to determine unknown parameters associated with IMc, including electrical, mechanical, and thermal. [19]. The use of MATLAB for parameter estimation of IMc is becoming increasingly popular due to its ease of use, flexibility, and accuracy. MATLAB provides tools for analyzing and estimating the parameters of IMc. The Simulink/Power System Blockset simulates steady-state behaviour, while the Motor Control Blockset estimates AC IMt parameters. The IMt Configuration block generates a configuration signal for updating parameters, allowing for efficient analysis and simulation of induction machine parameters. MATLAB is a powerful tool for estimating parameters in industrial applications, particularly for three-phase AC IMt. The Induction Machine Squirrel Cage block is useful for this purpose, allowing fundamental parameters to be expressed in per-unit or SI system units. The Motor Control Blockset offers parameter estimation blocks specifically designed for AC IMt, enhancing induction machine performance in various settings. [19, 20] [36].**Induction Machine Optimization:** Optimizing IMc is an important aspect of improving their performance. Optimization techniques can be used to determine the best design parameters for IMc, such as the number of turns in the stator and rotor windings, the size and shape of the rotor bars, and the air gap length. Objective functions are used to evaluate the performance of the IMc and can include parameters such as efficiency, torque, and power factor. By optimizing these parameters, IMc can be more efficient and reliable, leading to improved performance and reduced energy consumption. MATLAB is a powerful tool for optimizing IMc. It allows for the implementation of both static and dynamic modelling of three-phase IMc, which can help identify faults and improve performance [2]. MATLAB also provides a variety of optimization techniques, such as random restart local search optimization and fuzzy dynamic objective function control, which can be used to optimize IMc [37, 38]. Besides, MATLAB can be used to perform tests on IMc; one can conduct dc, no-load, and blocked-rotor tests to determine the equivalent circuit parameters and estimate the performance of machines. These tests are beneficial in identifying the machine's characteristics and offer valuable information for analysis and optimization [2, 3]. The optimization of IMc using MATLAB is an important area of research and development. Various control strategies, such as direct torque control and field-oriented control, can be implemented using MATLAB to improve the performance of IMc [1]. Using MATLAB to optimize IMc has the potential to improve energy efficiency and performance greatly. This can positively impact industries such as manufacturing, transportation, and renewable energy Fields [39].

Conclusion

This paper reviews enhancing the learning of IMc through MATLAB. However, modelling and controlling induction devices in MATLAB requires understanding their behaviors, which is difficult. Static and dynamic modelling of three-phase IMc and sensor less indirect field-oriented control (IFOC) with fuzzy logic are among the modelling and control methods. Multiple studies show that MATLAB/Simulink is useful for applying these strategies. Future IMc modelling and control research includes developing improved control algorithms, exploring new intelligent methodologies, and improving existing models to account for failures and high-performance scenarios. Induction machine testing and characterization require more investigation. Moreover, MATLAB-based IMc modelling and control is a fast-expanding field with great potential. MATLAB/Simulink helps researchers and engineers' model, simulate, and test IMc, improving application performance and efficiency.

References

1.   Le Roux, P.F. and M.K. Ngwenyama, *Static and Dynamic Simulation of an Induction Motor Using Matlab/Simulink.* Energies, 2022. **15**(10).

2.   Ayasun, S. and C.O. Nwankpa, *Induction Motor Tests Using MATLAB/Simulink and Their Integration Into Undergraduate Electric Machinery Courses.* IEEE Transactions on Education, 2005. **48**(1): p. 37-46.

3.   Ali, M.M.K.A.M. *Parameters Estimation Tests of Induction Machine Using Matlab/Simulink.* in *J. Phys.: Conf. Ser. 1973.* 2021. University of Babylon, Iraq.

4.   Saghafinia, A., *Teaching of Simulation an Adjustable Speed Drive of High Performance Induction Motor Using Matlab/Simulink,* in *Applications of Various Fuzzy Sliding Mode Controllers in Induction Motor Drives* A. Saghafinia, Editor. June 2016, Nova Science Publisher. p. 83-106.

5.   Saghafinia, A., et al., *Teaching of Simulation an Adjustable Speed Drive of Induction Motor Using MATLAB/Simulink in Advanced Electrical Machine Laboratory.* Procedia - Social and Behavioral Sciences, 2013. **103**: p. 912-921.

6.   Pratama, D.A., M. Anisah, and K.A. Setiyadi, *The Three Phase Induction Motor Test Using MATLAB 2021b/SIMULINK at Bukit*

Energi Servis Terpadu, Ltd. International Journal of Research in Vocational Studies (IJRVOCAS), 2022. **1**(4): p. 60-65.

7. Aktaibi, A., D.F.D. Ghanim, and M.A. Rahman. *Dynamic simulation of a three phase induction motor using Matlab Simulink.* 2011.

8. Al-suod, M.M.S. and A.O. Ushkarenka. *Analytical Representation of Control Processes of Induction Motor and Synchronous Generator in Power Plants.* 2016.

9. M. Covino, M.L.G.a.E.P., *Analysis of braking operations in present-day electric drives with asynchronous motors* in *1997 IEEE International Electric Machines and Drives Conference Record.* 1997: Milwaukee, WI, USA,. p. MB3/1.1-MB3/1.3.

10. Islam, S.M. and M.T. Iqbal, *Power Quality Estimation in a Remote Wind-Diesel Hybrid Power System in Cartwright, Labrador.* IU-Journal of Electrical & Electronics Engineering, 2011. **11**: p. 1319-1326.

11. Kumar, V.S. and H. Balaga, *Rotor Flux Estimation of Induction Motor Using Artificial Neural Networks.* Journal of Advance Research in Electrical & Electronics Engineering (ISSN: 2208-2395), 2015. **2**(11): p. 01-08.

12. NithinK, S., B.M. Jos, and M. Rafeek, *An Improved Method for Starting ofInduction Motor with Reduced TransientTorque Pulsations.* International Journal of Advanced Research in Electrical, Electronics and Instrumentation Energy, 2013. **2**: p. 462-470.

13. Tolbert, B.O.a.L.M., *Simulink implementation of induction machine model - a modular approach*, in *IEEE International Electric Machines*

and Drives Conference, 2003. IEMDC'03.* 2003: Madison, WI, USA.

14. Salim, S. and V. George, *Wide Speed Range Flux Estimator for Direct Vector Controlled Induction Motor.*, in *2013 International Conference on Control Communication and Computing (ICCC)* 2013: Thiruvananthapuram, India p. 373–77.

15. Purdue e-Pubs - Purdue University. (n.d.) Retrieved May 28, from docs.lib.purdue.edu

16. Al-Asady, H.A.-J. and H.F. Fakhruldeen, *Volts/Hz Control of Three Phase Induction Machine: Matlab Simulink*, in *J. Phys.: Conf. Ser. 1818 012089.* 2021.

17. Maurya, R. *Mathematical Modeling of Induction Motor . MATLAB Central File Exchange* 2023 [cited 2023 June 3].

18. M., S.S. and A.T. Patel, *Mathematical Modelling of an 3 Phase Induction Motor Using MATLAB/Simulink.* IJSRSET, 2016. **2**(3): p. 137-141.

19. Khalaf, M.M. and A.M. Ali, *Parameters Estimation Tests of Induction Machine Using Matlab/Simulink.* Journal of Physics: Conference Series, 2021. **1973**(1).

20. Salimin, R.H., et al., *Parameter Identification of Three-Phase Induction Motor using MATLAB-Simulink*, in *2013 IEEE 7th International Power Engineering and Optimization Conference (PEOC02013).* 2013: Langkawi. Malaysia.

21. Mathworks. *Induction Machine Control.* 2023 June 2 2023].

22.	*Induction Motor Tests Using MATLAB/Simulink. (n.d.) Retrieved May 28, 2023, from www.slideshare.net*

23.	Versele, C., O. Deblecker, and J. Lobry. *Implementation of a vector control scheme using dSPACE material for teaching induction motor drive and parameters identification.* in *18th International Conference on Electrical Machines.* 2008.

24.	Wade, S., M.W. Dunnigan, and B.W. Williams, *Modeling and simulation of induction machine vector control with rotor resistance identification.* IEEE Transactions on Power Electronics, 1997. **12**(3): p. 495–506.

25.	Moussavi, S.Z. and M. Fazly, *Learning improvement by using Matlab simulator in advanced electrical machinery laboratory.* Procedia - Social and Behavioral Sciences, 2010. **9**: p. 92-104.

26.	Diana, G. and R.G. Harley, *An Aid for Teaching Field Oriented Control Applied to Induction Machines.* IEEE Power Engineering Review, 1989. **9**(8): p. 69-70.

27.	Reddy, P.N., *Digital Simulation of Conventional Direct Torque Control of Induction Motor Drive with Reduced Flux and Torque Ripples.* International Journal Of Engineering Sciences & Research Technology 2013. **2**(10).

28.	Pereira, W.C.A., Aguiar, M. L., Paula, G. T., Monteiro, J. R. B. A., Bazan, G. H., Castoldi, M. F., & Sanches, D. S., *Virtual platform of Direct Torque Control of induction motor to assist in education of undergraduate students,* in *2015 IEEE 24th International Symposium on Industrial Electronics (ISIE).* 2015.

29. de Las Morenas, J., F. Moya-Fernandez, and J.A. Lopez-Gomez, *The Edge Application of Machine Learning Techniques for Fault Diagnosis in Electrical Machines.* Sensors (Basel), 2023. **23**(5).

30. Hammo, R., *Faults Identifification in Three-Phase Induction Motors Using Support Vector Machines,* in *College of Technology, Architecture and Applied Engineering.* 2014, Bowling Green State University.

31. Ali, M.Z., et al., *Machine Learning-Based Fault Diagnosis for Single- and Multi-Faults in Induction Motors Using Measured Stator Currents and Vibration Signals.* IEEE Transactions on Industry Applications, 2019. **55**(3): p. 2378-2391.

32. Bouzida, A., et al., *Fault Diagnosis in Industrial Induction Machines Through Discrete Wavelet Transform.* IEEE Transactions on Industrial Electronics, 2011. **58**(9): p. 4385-4395.

33. Salnikov, S., E. Solodkiy, and D. Vishnyakov, *Simulation of three-phase induction motor different bearing faults in Matlab Simulink environment,* in *J. Phys.: Conf. Ser. 1886 012009.* 2021.

34. Mohar, N.A., et al. *Fault Detection Analysis for Three Phase Induction Motor Drive System using Neural Network.* in *J. Phys.: Conf. Ser. 1878 012039.* 2021.

35. Gonzalez-Jimenez, D., et al., *Machine Learning-Based Fault Detection and Diagnosis of Faulty Power Connections of Induction Machines.* Energies, 2021. **14**(16).

36. Pandey, K. and P.H. Zope. *Estimating Parameters of a Three-Phase induction motor using Matlab/Simulink.* 2013.

37.   Selvam, P.P. and R. Narayanan, *Random restart local search optimization technique for sustainable energy-generating induction machine.* Computers & Electrical Engineering, 2019. **73**: p. 268-278.

38.   Yang, Z., et al., *Predictive current control of a bearingless induction motor model based on fuzzy dynamic objective function.* Transactions of the Institute of Measurement and Control, 2020. **42**(16): p. 3183-3195.

39.   Mohammad Noor, S.Z., M.K. Hamzah, and P.N.A. Megat Yunus, *Three phase induction motor analysis using MATLAB/GUIDE*, in *2013 IEEE 7th International Power Engineering and Optimization Conference (PEOCO)*. 2013. p. 161-166.

12. Role of Skill Development Centers in India for youth.

Dr. VajjalaNeelaveni

Assistant Professor, Department of Commerce and Management

GFGC, KR Puram, Bangalore, Karnataka, **India**

Abstract: The paper focuses on the various skills development programmes conducted by skill development centres in India for youth. India is rich in man power which is one of the strengths of the India. Entrepreneur is one of the factors of production. India needs the skilled man power for the development of nation. LPG phenomenon demands intensified the need of highly skilled workforce in developing nations. Demographically, India consists of 54% of population below the age of 25 years. Only about 2.5 million vocational training seats are available in the country whereas about 12.8 million persons enter the labour market every year. Many programmes have been undertaken by the government of India to develop the skilled youth in the country. Skills can be acquired through permanent learning process. The largest share of new jobs in India is likely to come from the unorganized sector that employs up to 93 per cent of the national workforce, but most of the training programmes cater to the needs of the organized sector.

Curriculum for skill development has to be reoriented on a continuing basis to meet the demands of the employers/industry and align it with the available self-employment opportunities. Accreditation and certification system has to be improved. In order to bridge the industry academia gap, NSDC has developed a unique model to integrate skill based trainings into the academic cycle of the Universities.

Keywords: Skill Development Centres, National Skill Development Corporation, Kaushal Kendra,DDU-GKY, RVTI and HTS.

Objectives of the Study

To identify the skilled man power to build developing nation;

To highlight the significant activities of skill development centres in India;

Methodology

The information collected from secondary sources such as magazines and relevant websites.

Introduction

Now days, skill India became a buzz word around the country. Skill is only the source to develop India. The graduate and post graduates can survived in the competitive world with some skills. The government identified the skills are important to the youth to compete in any field.

Skill training interventions raises confidence, improves productivity & competency of an individual through focused outcome based learning. Historically it is proved that mental and physical skilled people are employed in various professions. Idle youth is the biggest burden to the economy. The economy should focus on job creation and social security schemes for the youth. On the other hand, make it mandatory for the youth to take up jobs at least for certain fixed period to avail the social security benefits. The country's GDP also depends on the skills of entrepreneurial development. The practical skills are more needed than theoretical which cannot be known to implement properly. The student came out of the campus with some skills by doing several certificate courses, they can get alternative jobs on their acquired skills or else they can establish a star - ups. Education system needs some changes in Indiatocaptureemployment opportunities. The competent students may not get the suitable jobs at the right time without having required soft skills. Particularly in India, skill oriented course to be introduced to get the efficient human resources in the higher education. Various skill development centers have been established and emerged to fulfill the required skills among the youth. Fortunately, the present government is giving importance to develop the skills among youth. Students are also should identify the importance of required skill and have to be acquired them. India has large young population, only 5% of the Indian labour force in the age group of 20-24 years has obtained vocational skills through formal means whereas the percentage in industrialized countries varies between 60 % and 96%. About 63% of the school students drop out at different stages before reaching SSC. Only about 2.5 million vocational training seats are available in the country whereas about 12.8 million persons enter the labour market every year. Even out of these training places, very few are available for early school dropouts. This indicates that a large number of school drop outs do not have access to skill development for improving their employability. The educational entry requirements and long duration of courses of the formal training system are some of the impediments for a person of low educational attainment to acquire skills for their livelihood. The largest share of new jobs in India is likely to come from the unorganized sector that employs up to 93 per cent of the national workforce, but most of the training programmes cater to the needs of the organized sector. In 2014, Skill Development started from the government under the visionary leadership of Honorable Prime Minister. He encouraged Skill India mission and also formed Ministry of Skill Development & Entrepreneurship Skill (MSDE) to coordinate all skill development activities, capacity & technical/ vocational training framework building, assessments framework. The Ministry is dedicated to be skilled 400 million workforce by 2022.Today, a large section of India's labor force has outdated skills. With current and expected economic growth, this challenge is going to only increase further, since more than 75% of new job opportunities are expected to be 'skill-based. Corporate educational institutions, non-government organizations, government, academic institutions, and society would help in the development of skills of the youths so that better results are achieved in the shortest time. Every job aspirant should be given training in soft skills to lead a proper and decent life. **Skill India:** 'Skill India' is a dream project of Honorable Prime Minister, NarendraModi and the work to launch this programme has already been initiated. The main goal is to create opportunities and scope for the development of the talents of the Indian youth and to develop more of those sectors which have already been put under skill development for the last so many years and also to identify new sectors for skill development. The new programme aims at providing training and skill development to 500 million youth of country by 2020, covering each and every village.Corporate educational institutions, Non-government organizations and society should help the skill development to be the main focus for the economy to grow. The early development of skill in youth, as early as school level is very important to groom them for proper job opportunities. The training should be given in such a way that the youth is suitable for jobs in any part of the world. Today the world needs to give importance for all the jobs equally so that there will be balanced growth in all the sectors. Every job aspirant whether he/sh is white collar or blue collar should be given training in soft skills. Soft skills are needed in every area, grooming, etiquette, hygiene, time management, safety, tolerance levels to lead a proper and decent life. India is one of the few countries in the world where the working age population will be far in excess of those dependent on them and as per the World Bank. The emphasis is to skill the youth in such a way that they get employment and also improve entrepreneurship. It Provides training, support and guidance for all occupations that were of traditional type like carpenters, cobblers, welders, blacksmiths, masons, nurses, tailors, weavers etc. More emphasis will be given on new areas like real estate, construction, transportation, textile, gem industry, jewellery designing, banking, tourism and various other sectors, where skill development is inadequate or nil. The training programmes would be on the lines of international level so that the youth of our country can not meet the domestic demands and of other countries like the US, Japan, China, Germany, Russia and West Asia. Another remarkable feature of the 'Skill India' programme would be to create a hallmark called 'Rural India Skill', so as to standardise and certify the training process. Tailor-made, need-based programmes would be initiated for specific age groups which can be like language and communication skills, positive thinking skills, personality development skills, management skills, behavioural skills and employability skills. The course methodology of 'Skill India' would be innovative, which would include games, group discussions, brainstorming sessions, practical experiences, case studies etc. Curriculum for skill development has to be reoriented on a continuing basis to meet the demands of the employers/industry and align it with the available self-employment opportunities. Accreditation and certification system has to be improved. There is a need to establish an institutional mechanism for providing access to information on skill inventory and skill maps on a real time basis. A sectoral-approach is required for the purpose with special emphasis on those sectors that have high employment potential.

Skill Development Centres can be established in existing education and training institutions. This would ensure huge saving in cost and time. A system of funding poor people for skill development through direct financial aid or loan also needs to be put in place. Apprenticeship training as another mode for on-job training has to be re-modeled to make it more effective and up-scaled significantly.

The required skills to be learned from skill development centres

1. Language Skills
2. Communication Skills
3. Positive thinking Skills
4. Life Skills
5. Personality Development skills
6. Mentoring skills
7. Management skills
8. Job skills
9. Behavioural skills
10. Tolerance skills
11. Employability skills

National Skill Development Corporation (NSDC): The National Skill Development Corporation India (NSDC) was setup as one of its kind, Public Private Partnership Company with the primary motive of speed up of the skills in India. The AISECT-NSDC certified centers, a national skill development agency have been established with an aim to provide youth with university certified undergraduate, postgraduate, certificate and diploma courses along with low cost, high quality teaching.

The objectives of the NSDC

1. Upgrade skills to international standards through significant industry involvement and develop necessary frameworks for standards, curriculum and quality assurance

2. Enhance, support and coordinate private sector initiatives for skill development through appropriate Public-Private Partnership (PPP) models.

3. Play the role of a "market-maker" by bringing financing, particularly in sectors where market mechanisms are ineffective.

Functions of NSDC: Private Sector – Areas of partnerships include awareness building, capacity creation, loan financing, creation and operations of Sector Skill Councils, assessment leading to certification, employment generation, Corporate Social Responsibility, World Skills competitions and participation in Special Initiatives like Udaan focused on J&K. International Engagement – consultancy in Investments, technical assistance, transnational standards, overseas jobs and other areas. Central Ministries – Participation in flagship programmes like

Make in India, Swachh Bharat, Pradhan Mantri Jan DhanYojana, Smart City, Digital India and Namami Ganga.

State Governments – Development of programs and schemes, alignment to NSQF and capacity building, operationalization of program, capacity building efforts among others. University/School systems – Vocational education through specific training programs, evolution of credit framework, entrepreneur development, etc. Non-profit organizations – Capacity building of marginalized and special groups, development of livelihood, self-employment and entrepreneurship programs. Innovation – Support to early-stage social entrepreneurs working on innovative business models to address gaps in the skilling ecosystem, including programs for persons with disability. *Achievements:* From NSDC over 5.2 million students trained by235 private sector partnerships organizations for training and capacity building. Eachorganization has trained at least 50,000 persons over 10-yearsof time.Totally 38 Sector Skill Councils (SSC) approved in services, manufacturing, agriculture & allied services, and informal sectors. Sectors include 19 of 20 high priority sectors identified by the government and 25 of the sectors under Make in India initiative. There are 1386 qualification packs with 6,744 unique National Occupational Standards (NOS). These have been validated by over 1000 companies. There are Vocational training centres introduced in 10 States, covering above 2400 schools, 2 Boards, benefitting over 2.5 lakh students. Curriculum based on National Occupational Standards (NOS) and SSC certification. NSDC is working with 21 universities, Community Colleges under UGC/AICTE for alignment of education and training to NSQF. It involves Skill Development Management System (SDMS) with 1400 training partners, 28,179 training centres, 16,479 trainers, 20 Job portals, 77 assessment agencies and 4,983 empanelled assessors. Hosting infrastructure certified by ISO 20000/27000 supported by dedicated personnel. The National Skill Development Corporation (NSDC) is a Public Private Partnership, set up to secure the setting-up of large scale, profit sustainable vocational institutions in the country by encouraging private sector participation and providing low-cost funding for training capacity. The programmes are provided by the AISECT in several disciplines like AISECT Academy for IT & Management, Hardware & Networking, Teacher Training, Livelihood & Vocational Training, Insurance, Banking & Finance, Agriculture, Textile Training, Fire Safety & Security, Auto Skills, Telecom Skills, and Retail Management. NSDC is working with 21 Universities, UGC and AICTE catering to more than 1200 colleges and 400 community colleges across the country. Some of the organization include:

1. 663 colleges and 57 autonomous institutionsof SavithribaiPhule university of Pune.
2. 67 colleges including NCWEB and SOL of university of Delhi.
3. 150 community colleges and 127 colleges for B.Voc and Degree programmes from University Grants Commission.

4. 155 learning resource centres and 204 community colleges from Tamil Nadu Open University (TNOU).

5. 4 colleges from Centurion University.

6. Haryana and Punjab Universities

7. 100 Community Colleges from AICTE

Pradhan MantriKaushal Kendra: Pradhan Mantri Kaushal Kendra (PMKK) inaugurated by Minister of State for Kill Development and Entrepreneurship, in Delhi on 27th September, 2017. The center is located at Kirari in North-west Delhi and is spread over an areas of 20000 Sq.ft. IACM smart learn limited a Delhi based training company, will operating the training partner under skill India. The newly launched PMKK has the target to impart training to 2260 candidates by financial year 2018 in the following eight job roles –Beauty, plumbing, retail, construction, Electronic, furniture fittings, domestic workers and security. Pradhan Mantri Kaushal Vikas Yojana (PMKVY) is the flagship scheme of the Ministry of Skill Development & Entrepreneurship (MSDE). The objective of this Skill Certification Scheme is to enable a large number of Indian youth to take up industry-relevant skill training that will help them in securing a better livelihood. Individuals with prior learning experience or skills will also be assessed and certified under Recognition of Prior Learning (RPL). Ministry of Skill Development and Entrepreneurship (MSDE), through the National Skill Development Corporation (NSDC), has pioneered d to establish Pradhan MantriKaushal Kendra (PMKK) in every district of

India **DeenDayalUpadhyaGrameenKaushalya Yojana (DDU – GKY)** DDU-GKY is a prominent government youth employment scheme in India. The scheme launched 25th September, 2014 on the occasion of 98th birth anniversary of Pandit Deendayal Upadhyaya. The vision of DDU-GKY is to "transform rural poor youth into an economically independent and globally relevant workforce." The Scheme is the skill development and placement programme initiated by the Ministry of Rural Development (MORD), government of India. The target group of the scheme is youth, under the age group of 15-35 years. It is a part of the National Rural Livelihood Mission (NRLM) having aim of adding diversifies to the incomes of rural poor families and caters to the career aspirations of rural youth. Nearly, Rs. 15 crore fund allocated for the programme. DDU-GKY is working in 21 states and union territories, across 568 districts impacting youth from over 6,215 blocks. Through this programme nearly 2.7 lakhs candidates placed. The scheme focuses on catering to the occupational aspirations of rural youth and enhancing their skills for wage employment. Implementation of DDU_GKY involves state governments' technical support agencies like the National Institute of Rural Development and Panchayathi Raj (NIRD & PR) The Standard Operating Procedures (SOP) to be followed for the implementation of projects under DDU_GKY. Two categories of courses are available on SOP e-learning portal.

Professional	Master Trainer

Operations	Operations
Finance	Finance
Comprehensive	comprehensive

The portal allows the candidates to learn at their own pace. The duration of course depends entirely on candidates. All the materials required to the course are provided under the 'Resources tab of the SOP e-learning portal. A certificate will be awarded to everyone who clears the final certification examination. The certificate is valid for one year, the time of renewal of certificate will be notified to the candidates. The candidates can register their names from the website of DDU-GKY. **Regional Vocational Training Institute (RVTI):** Recently Vice President of India lays Foundation Stone for the first Regional Vocational Training Institute (RVTI) in Hyderabad. RVTI will help boost women empowerment.RVTI campus will be spread over 4 acre of land; around 1000 candidates will be trained annually. 5 new PMKKs e-launched across Uttar Pradesh and Madhya Pradesh. The competent authority has approved the expenditure of Rs. 19, 95, 90,000/- for construction of building for RVTI-Hyderabad. RVTI campus will be spread over approximately four acres of land with the campus of ATI-Vidyanagar, Hyderabad. After getting its new permanent building, around 1000 trainees would be trained here every year. This RVTI is expected to cater to the vocational training needs of 480 women annually in regular courses under Craftsmen Training Scheme (CTS) and Craft Instructor's Training Scheme (CITS) and around same number of candidates would be trained under various short term courses in skill areas having a high market demand. The courses identified to be run in this Institute are Fashion Design & Technology, Architectural Assistantship, Cosmetology, Front Office Assistant, Secretarial Practice (English), Food and Beverages Service Assistant. During the event, honorable Vice President handed over Recognition of Prior Learning (RPL) certificates and placement certificates to candidates who got placed in companies like Tech Mahindra & Accenture. Recognition of Prior Learning recognizes an individual's skills through certification and enhancing their career opportunities. Regional Vocational Training Institute (RVTI) for women are located in various places such as Mohali, Shimla, Jaipur, Mumbai, Vadodara, Thiruvanthapuram,Tura, Indore, Panipat, Bangalore, Allahabad and Kolkata. **Hi-tech Training Schemes (HTS):** Hi-tech Training Scheme is one of the schemes of the erstwhile World Bank assisted Vocational Training Project. The scheme is now being continued for implementation with Government of India funding. It consists of State institutes in which 10 ITIs are working. Central institutes located at Bangalore. The objective of the Hi-tech scheme is to produce trained personnel with the range of skills necessary to meet the requirements of industry, commerce and domestic consumers in the application of electronics, computer and the modern production system. Short-term Courses of 2-3 weeks duration in the following Hi-tech areas are envisaged / being implemented in the ATIs / ATI-EPI for the industries / Public Sector Undertakings / Government organizations / Trainers from the institutes/industries etc.

Conclusion: Now – a - days, acquiring skills is the source to develop India. The graduates and post graduates can be survived in the competitive world with the best skills. Indian education sector has seen rapid growth in number of institutions and students over last few decades. The skill development initiative scheme is 100 % centrally sponsored scheme. As per UGC report, in 1950-51 there were approximately 750 colleges affiliated to 30 universities, which has grown to over 727 universities, 35000 colleges & 13000 standalone Institutions in 2014-15. In today's world of globalization, skill training is an integral component of increasing efficiency & productivity for sound economic development of any economy. Generally universities revise the curriculums of courses periodically through the board of studies (BOS). Accordingly, Curriculums for skill development has to be reoriented on a continuing basis to meet the demands of the employers/industry and align it with the available self-employment opportunities.

13. Phytochemicals in structure-based drug discovery for cancer treatment- a review of the evidence.

Dr Pooja R

Department of Biotechnology, Surana College Autonomous, South end road, Bangalore, Karnataka, **India**

Abstract: Cancer has been one of the leading causes of death worldwide for many years, owing to conventional medicines that have faced numerous hurdles, including multidrug resistance, side effects, and cost-effectiveness. As a result, there is a demand for complementary alternative medicine (CAM) for cancer treatment that improves prognosis while remaining cheap. Plant-based phytochemicals have recently been discovered to have various therapeutic properties such as anticancer, antioxidation, anti-inflammation, immunomodulation, and so on. Taxol, vinca alkaloids (vincristine and vinblastine), and podophyllotoxin are examples of common phytochemicals that have anticancer properties by affecting cancer growth and progression molecular pathways. Furthermore, pioneer phytochemicals have served as models for the design and development of novel anticancer drugs with enhanced pharmacokinetics and pharmacodynamics. The FDA-approved anticancer medicines doxorubicin and idarubicin, were created by utilising an anthraquinone scaffold and starting ingredients. As a result, the purpose of this chapter is to describe the phytochemicals, modes of action, and structure employed as scaffolds for novel drug development.

Keywords: Cancer; Phytochemicals; and Antitumor activity.

Introduction: According to WHO 2011, cancer is one of the top causes of death worldwide, accounting for 7.9 million deaths in 2007. It is also one of the most obnoxious diseases in emerging and impoverished countries because to a lack of effective remedies and care [1]. Every year, around 11 million people are diagnosed with cancer, and millions of people die from it. According to several studies, various forms of cancer (stomach, lungs, cervix uteri, liver, breasts, and colorectal) account for approximately 13% of all deaths each year [2]. Significant attempts were made to identify the epitome technique for cancer medication study, which eventually emerged as the major thrust of study for multiple eras [3]. According to various studies, women have a greater cancer mortality rate than men. Males had a greater occurrence incidence for prostate (15.0%), lung (16.7%), stomach (8.5%), colorectum (10.0%), and liver (7.5%) cancers, whereas females had an elevated occurrence rate of roughly 25.2% - breast, lung -8.7%, colorectum -9.2%, stomach -4.8%, and cervix -7.9% [4]. Malignancy is defined as the uncontrollable proliferation of abnormal cells everywhere in the human body. Malignant cells or tumour cells are the designation given to newly formed abnormal cells. Malignant cells have the ability to infect healthy tissues. Several cancers and the irregular units that form the malignant tissue are then identified by the specific tissue where these irregular cells are created (for instance, cervical cancer, breast cancer, and colorectal cancer). These cells frequently segregate from the initial source of mass cells, proliferate through multiple channels (blood and lymph), and reside in new organs from which they can repeat the cycle. This progression from one section of the body to another is referred to as metastatic disease or metastasis.

Historical perceptions of phytochemicals as anti-cancer agents:

Cancer treatment relies heavily on biological products. Particularly Secondary metabolites are essential for the efficient operation of the human body's inherent mechanisms. Several studies have found that phytochemicals decrease highly expressed proteins, amino acids, hormones, and enzymes in a variety of ways. Plant-derived chemicals are being used to inhibit cancerous tumours and reduce treatment resistance. These can control the expression of genetic abnormalities, change in genetic expression of certain genes such as p21 and p53 during mitosis or meiosis, and even carry out DNA repair mechanisms. Scientists are developing semi-synthetic analogues with enhanced pharmacological qualities with the awareness of biological molecules to cure malignancies, Chemical lead molecules, as well as medicinal plants and other natural resources [5]. As a result of significant research by chemists and pharmacologists, several phytotherapeutic procedures have been described [6]. These phytochemicals help to regulate the pace at which protective enzymes are generated, and they have also been shown to have antioxidant and relative oxygen-producing properties by influencing multiple pathways [7]. Some cytotoxic medications inhibit angiogenesis while causing low damage. Surveys on new plants will lead to the development of novel anticancer medications, the success of which will be monumental [8]. Meanwhile, specialised target drugs have been designed to attack tumor-related proteins, and they are the foundation of precise medicine; biological molecules derived from plant resources represent a valuable source for specific remedies because they affect multiple targets at once; and pharmacological combinations behave in a multi-specific manner.

Enhancement and Application of Synthetic Analogs for Plant-Derived Compounds

One of the key limiting aspects of secondary metabolites extracted from plants is their low solubility or insufficient bioavailability, which prevents their application in clinical trials. The use of semi-synthetic or synthetic analogues for identifying plant-derived substances has been applied as the ultimate solution to this challenge [9]. Morphine has been transformed to morphine-6-glucuronide, for example, to boost medicinal properties. Paclitaxel (Taxols) and its analogues docetaxel (Taxoteres) and cabazitaxel (Jevtanas) are two examples of scientifically proven synthetic anticancer analogues derived from plants;

camptothecin and its analogues belotecan (Camptobells), irinotecan (Camptosars); vinblastine (Vumons).

Classification of Phytochemicals:

In pharmaceutical treatment, natural compounds play a key role, particularly in the case of antitumor drugs. Beneficial component classes such as steroids, fatty acids, glycosides, terpenes, flavonoids, alkaloids, tannins, and phenolics have been identified in ethnomedicine plant analysis. [10]. They are responsible for enhancing DNA repair pathways and directly affect the central hallmark of cancer progression and metastasis. The pure, chemically well-defined drugs obtained naturally have been used in before being used as a drug, advanced medicines are either transformed directly or through chemical processes. Plant alkaloids have an impact on a variety of underlying signalling mechanisms. [11]., Some alkaloids (capsaicin, piperine) appear to enhance tumour growth and metastasis, or to act as co-carcinogens, whereas others appear to be genotoxic (caffeine, sanguinarine, harmine). Consequently, cancer-curing medications must not be genotoxic or carcinogenic. Caffeine has been known to help the development of mammary glands in addition to DNA damaging chemicals that induce cancer. This finding could be regarded as a proliferative impact, which is undesirable in anticancer medicines as well. Phenols, flavonoids, and their derivatives, stilbenes and lignans are among the many antioxidants that fall under this category. Polyphenols have various targets in carcinogenesis, drug and radiation resistance mechanisms, tumour cell proliferation and apoptosis, inflammation, invasive dissemination, angiogenesis, and drug and radiation resistance and processes. [12]. Polyphenols inhibit platelet activation, capillary permeability, lipid peroxidation, and enzyme systems like lipoxygenase, among other biological activities. The following phytochemicals have been found to be useful in the chemoprevention of cervical cancer in experimental investigations.

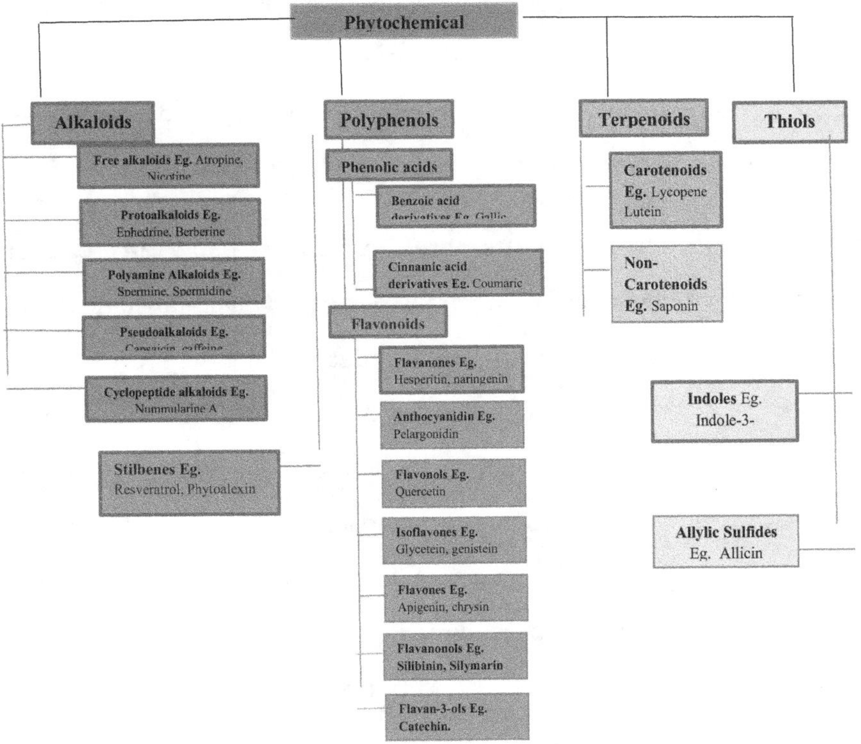

Terpenoids are categorized into various groups, depending on the amount of building unitss, such as monoterpenes (e.g., carvone, perillyl alcohol , limonene-*d*, and geraniol), diterpenes (e.g. *trans*-retinoic acid and retinol), triterpenes [e.g., lupeol, ursolic acid, oleanic acid, and betulinic acid], and tetraterpenes (e.g., lutein, α-carotene, , and β-carotene lycopene [13]. Many triterpenoids have been found to inhibit the growth of a wide variety of cell types while causing no harm to healthy cells [14]. Thiols are a group of highly reactive chemicals that play an important role in maintaining the cellular redox equilibrium [15]. The thiol moiety is one of the cell's most powerful nucleophilic groups. It has taken part in a variety of biochemical process., where thiol-containing compound have an important function in maintaining intracellular redox balance, commonly known as oxidative stress, Apoptosis is another well-known role for the phytochemical [16].

Phytochemicals from plants are currently used in cancer treatment: Phytochemicals, or naturally occurring plant substances, are employed in the development of new drugs and in the treatment of cancer. It is classified based on its chemical structure [17]. Some examples include taxol analogues, vinca alkaloids such as podophyllotoxin, vinblastine, vincristine, and analogues. Phytochemicals typically interact by influencing the molecular pathways of cancer growth and progression. Individual techniques include increased antioxidant levels, suppression of proliferation, encouragement of cell cycle arrest and apoptosis, carcinogen inactivation, and immune system regulation [18]. They perform a wide range of activities on a wide range of biological targets and signalling pathways, including membrane receptors [19]. Phytochemicals' anticancer properties have been studied in vitro and in vivo. They have separate and overlapping processes that scavenge free radicals in order to cut down carcinogenic activity [20]. Figure 1 depicts the chemical structures of numerous chemicals produced from plants that are utilised to treat cancer in this chapter, Since ancient times, nature has provided medicinal herbs, which have played a critical role in human progress. Phytochemicals derived from various portions of medicinal plants have a variety of therapeutic properties, including inhibiting development and progression of cancer. Secondary metabolites play a key role in hindering signalling pathways of such as cyclooxygenase activity, topoisomerase enzyme, and CDK4 kinases activity, as well as DNA repair mechanisms activation and enhancing protective enzymes formation such as capase-3, resulting in strong anticancerous effect.

Different strategies for the development of anticancer phytochemicals: Plants with medicinal properties play a important role especially in the field of pharmaceutical industry for the treatment of various diseases. They might vary with longitude, latitude, altitude, climate, age, and seasonal diversity from species to species. Each plant parts contain many numbers of pharmacological functions. These bioactive phytochemicals are majorly used as anticancer drugs. Phytochemical purification involves many stages i.e., isolation assays, bioassay-guided fractionation and combinatorial chemistry. For the separation of bioactive compounds, various analytical techniques have been used. The process starts with natural extracts analysis by using dry or wet plant material by confirming the biological activity. Then, followed by fractionation of active plant extracts, tested for biological activities and also various analytical techniques such as GCMS, LCMS, HPLC, TLC, UV-vis, NMR and FTIR, etc. are used for the separation of active compounds. For some of the polarity orders, different solvents are used. Sephadex, Superdex, Silica, or any other suitable matrix can be used for fractionation purposes. Even some of the dyeing agents are used for the detection of natural compounds in endangered species or even in medicinal plants (e.g. Vanilline sulfuric acid). These protocols may vary from species to species. However quality, purity, and quantity of the bioactive compounds must be very high and this can be done by using the highly pure solvents, and matrices, and also with careful handling. Once the purification is done for the phytochemicals, they should undergo in-vitro or in-vivo anticancer effects. If the compound showed anticancer activity then it has to undergo various tests such as pharmacokinetics, pharmacodynamics, immunogenicity, metabolic fate, biosafety, side effects, drug interactions, dose concentration , etc. for future drug design. A detailed schematic diagram is shown below of bioactive compound synthesis, optimization, characterization, testing, and potential application as a cancer therapeutic agent is shown in Figure 2. Detailed scheme of anticancer phytochemical synthesis, optimization, characterization and prospective use as cancer therapeutic agent.

Bioactive Compounds and Their Anticancer Functions: Previously, many varieties of food which includes vegetables and fruits were determined to have a great impact on human health. Bioactive compounds contain many secondary metabolite and help in the prevention of diseases thereby protecting human health [40].

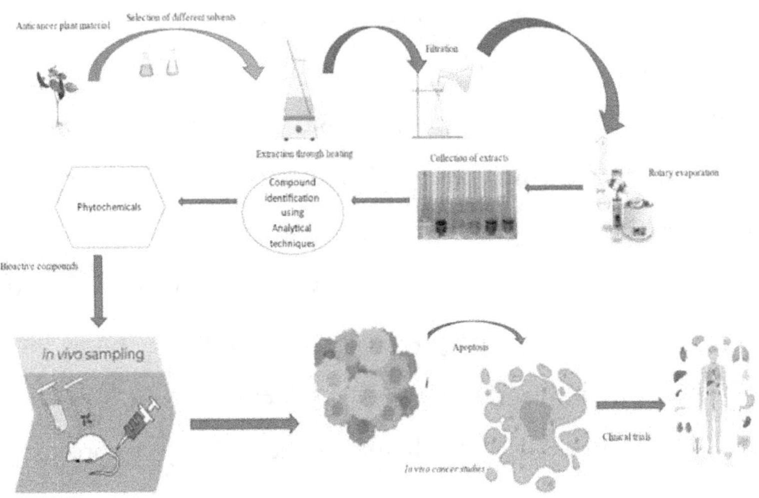

Most of the secondary metabolites vary in their structure and functions and also have the potential to act as chemotherapeutic and chemopreventive agents, especially in the treatment of cancer [41-44]. The bioactive compounds that will be discussed and the food sources of these compounds Table 2 [45].

Conclusion: Cancer is the leading cause of death worldwide. It is necessary to develop new drugs for the prevention of cancer disease.

Around 60% of the current anticancer drugs are derived from the natural sources. Nature continues to be an abundant source of biologically active and diverse chemotypes, and while relatively only a few of the actual isolated natural products are developed into clinically effective drugs in their own right. Most of the bioactive compounds are identified as chemo preventive and chemotherapeutic agents in the treatment of cancer. These unique molecules often serve as models for the preparation of more efficacious analogs and prodrugs through the chemical methodology, such as total or combinatorial (parallel) synthesis, or the manipulation of biosynthetic pathways. A novel approach for the drug discovery was ethnopharmacological knowledge which is supported by the broad spectrum which involves pharmacology, biochemistry, medicinal chemistry, molecular, and cellular biology including natural product chemistry which is essential to harvest the potentials of phytochemicals. In addition, improvements in the formulation may result in the more effective administration of the drug to patients, or conjugation of toxic natural molecules to monoclonal antibodies or polymeric carriers specifically targeting epitopes on tumors of interest can lead to the development of efficacious targeted therapies. The essential role played by natural products in the discovery and development of novel anticancer agents, and the importance of multidisciplinary collaboration in the optimization of novel molecular drugs from natural product sources have been extensively reviewed.

References

1. George, B.P., R. Chandran, and H. Abrahamse, *Role of phytochemicals in cancer chemoprevention: insights.* Antioxidants, 2021. 10(9): p. 1455.

2. Omara, T., et al., *Medicinal plants used in traditional management of cancer in Uganda: a review of ethnobotanical surveys, phytochemistry, and anticancer studies.* Evidence-Based Complementary and Alternative Medicine, 2020. 2020.

3. GOLDIN, A., *Historical development and current strategy of the National Cancer Institute drug development program.* Methods in cancer research, 1979. 16: p. 165-245.

4. Alves-Silva, J.M., et al., *North African medicinal plants traditionally used in cancer therapy.* Frontiers in pharmacology, 2017. 8: p. 383.

5. Efferth, T., *Molecular pharmacology and pharmacogenomics of artemisinin and its derivatives in cancer cells.* Current drug targets, 2006. 7(4): p. 407-421.

6. Jiang, P., et al., *Lactisole interacts with the transmembrane domains of human T1R3 to inhibit sweet taste.* Journal of Biological Chemistry, 2005. 280(15): p. 15238-15246.

7. Nasri, H., *Toxicity and safety of medicinal plants.* Journal of HerbMed Pharmacology, 2013. 2. 16.

8. Bahmani, M., et al., *Cancer phytotherapy: Recent views on the role of antioxidant and angiogenesis activities.* Journal of evidence-based complementary & alternative medicine, 2017. 22(2): p. 299-309.

9. Fridlender, M., Y. Kapulnik, and H. Koltai, *Plant derived substances with anti-cancer activity: from folklore to practice.* Frontiers in plant science, 2015. 6: p. 799.

10. Rex, J., N. Muthukumar, and P. Selvakumar, *Phytochemicals as a potential source for anti-microbial, anti-oxidant and wound healing-a review.* MOJ Biorg Org Chem, 2018. 2(2): p. 61-70.

11. Efferth, T. and F. Oesch. *Repurposing of plant alkaloids for cancer therapy: Pharmacology and toxicology.* in *Seminars in Cancer Biology.* 2021. Elsevier.

12. Asensi, M., et al., *Natural polyphenols in cancer therapy.* Critical reviews in clinical laboratory sciences, 2011. 48(5-6): p. 197-216.

13. Rabi, T. and A. Bishayee, *Terpenoids and breast cancer chemoprevention.* Breast cancer research and treatment, 2009. 115(2): p. 223-239.

14. Setzer, W. and M. Setzer, *Plant-derived triterpenoids as potential antineoplastic agents.* Mini reviews in medicinal chemistry, 2003. 3(6): p. 540-556.

15. Pivato, M., M. Fabrega-Prats, and A. Masi, *Low-molecular-weight thiols in plants: Functional and analytical implications.* Archives of biochemistry and biophysics, 2014. 560: p. 83-99.

16. Zhang, S., C.-N. Ong, and H.-M. Shen, *Critical roles of intracellular thiols and calcium in parthenolide-induced apoptosis in human colorectal cancer cells.* Cancer letters, 2004. 208(2): p. 143-153.

17. Kapinova, A., et al., *Are plant-based functional foods better choice against cancer than single phytochemicals? A critical review of current*

breast cancer research. Biomedicine & Pharmacotherapy, 2017. 96: p. 1465-1477.

18. Choudhari, A.S., et al., *Phytochemicals in cancer treatment: From preclinical studies to clinical practice*. Frontiers in pharmacology, 2020: p. 1614.

19. Deng, Q.-P., et al., *Effects of glycyrrhizin in a mouse model of lung adenocarcinoma*. Cellular Physiology and Biochemistry, 2017. 41(4): p. 1383-1392.

20. E. Ooko, O. Kadioglu, H.J. Greten, T. Efferth. Pharmacogenomic characterization and isobologram analysis of the combination of ascorbic acid and curcumin-two main metabolites of *Curcuma longa*-in cancer cells. Front Pharmacol (2017), 10.3389/fphar.2017.00038

21. T. Efferth. From ancient herb to versatile, modern drug: *Artemisia* annua and artemisinin for cancer therapy. Semin Canc Biol (2017), 10.1016/j.semcancer.2017.02.009

22. L.Q. Zhang, R.W. Lv, X.D. Qu, X.J. Chen, H.S. Lu, Y. Wang. Aloesin suppresses cell growth and metastasis in ovarian cancer SKOV3 cells through the inhibition of the MAPK signaling pathway. Anal Cell Pathol, 2017 (2017), pp. 1-6.

23. J. Bhandari, B. Muhammad, P. Thapa, B.G. Shrestha. Study of phytochemical, anti-microbial, anti-oxidant, and anti-cancer properties of *Allium wallichii*. BMC Compl Altern Med, 17 (1) (2017), p. 102.

24. N.A. Jaradat, R. Al-Ramahi, A.N. Zaid, O.I. Ayesh, A.M. Eid. Ethnopharmacological survey of herbal remedies used for treatment of various types of cancer and their methods of preparations in the West Bank-Palestine. BMC Comp Altern Med, 16 (1) (2016), p. 93

25. M.A. Beg, U.V. Teotia, S. Farooq. *In vitro* antibacterial and anticancer activity of Ziziphus. J Med Plants, 4 (5) (2016), pp. 230-233.

26. M. Kumari, B. Pattnaik, S.Y. Rajan, S. Shrikant, S.U. Surendra. EGCG-A Promis anti-cancer Phytochem, 3 (2) (2017), pp. 8-10.

27. S. Amin, H.K. Barkatullah. Pharmacology of *Xanthium* species. A review. J Phytopharm, 5 (2016), pp. 126-127

28. S.Q. Pang, G.Q. Wang, J.S. Lin, Y. Diao, R.A. Xu. Cytotoxic activity of the alkaloids from Broussonetia papyrifera fruits. Pharm Biol, 52 (2014), pp. 1315-1319.

29. A. Mehdad, G. Brumana, A.A. Souza, J.A. Barbosa, M.M. Ventu ra, S.M. de Freitas, *et al.*A Bowman-Birk inhibitor induces apoptosis in human breast adenocarcinoma through mitochondrial impairment and oxidative damage following proteasome 20S inhibition. Cell Death Dis, 2 (2016), p. 15067.

30. N.A. Jaradat, R. Al-Ramahi, A.N. Zaid, O.I. Ayesh, A.M. Eid. Ethnopharmacological survey of herbal remedies used for treatment of various types of cancer and their methods of preparations in the West Bank-Palestine. BMC Comp Altern Med, 16 (1) (2016), p. 93

31. C.C. Tsai, T.W. Chuang, L.J. Chen, H.S. Niu, K.M. Chung, J.T. Cheng, *et al.* Increase in apoptosis by combination of metformin with silibinin in human colorectal cancer cells. World J Gastroenterol, 21 (2015), pp. 4169-4177

32. Y. Wang, C. Hong, C. Zhou, D. Xu, H.B. Qu. Screening antitumor compounds psoralen and isopsoralen from *Psoralea corylifolia* L. seeds. EvidBased Comp Altern. Med, 2011 (2011), Article 363052, 10.1093/ecam/nen087.

33. W. Bhouri, J. Boubaker, I. Skandrani, K. Ghedira, L.C. Ghedira. Investigation of the apoptotic way induced by digallic acid in human lymphoblastoid TK6 cells. Cancer Cell Int, 12 (2012), p. 26.

34. L. Wang, G.F. Xu, X.X. Liu, A.X. Chang, M.L. Xu, A.K. Ghime ray, *et al. In vitro* antioxidant properties and induced G_2/M arrest in HT-29 cells of dichloromethane fraction from *Liriodendron tulipifera.* J Med Plants Res, 6 (2012), pp. 424-432.

35. R. Hoshyar, H. Mollaei. A comprehensive review on anticancer mechanisms of the main carotenoid of saffron, crocin. J Pharm Pharmacol (2017), 10.1111/jphp.12776.

36. M. Hu, L. Xu, L. Yin, Y. Qi, H. Li, Y. Xu, *et al.* Cytotoxicity of dioscin in human gastric carcinoma cells through death receptor and mitochondrial pathways. J Appl Toxicol, 33 (2013), pp. 712-722.

37. H. Lauritano, J.H. Andersen, E. Hansen, M. Albrigtsen, L. Escal era, F. Esposit, *et al.* Bioactivity screening of microalgae for

antioxidant, anti-inflammatory, anticancer, anti-diabetes, and antibacterial activities. Front Mar Sci, 3 (2016), p. 68.

38. Newman, D. J., G. M. Cragg. 2012. Natural products as sources of new drugs over the 30 years from 1981 to 2010. *J Nat Prod; 75:311-335.*

39. Lee, W.-L., J.-Y. Huang, and L.-F. Shyur, *Phytoagents for cancer management: regulation of nucleic acid oxidation, ROS, and related mechanisms.* Oxidative medicine and cellular longevity, 2013. 2013.

40. Bandopadhyaya, S., Ramakrishnan, M., Thylur, R. P., and Shivanna, Y. (2015). *In-vitro* evaluation of plant extracts against colorectal cancer using HCT 116 cell line. *Int. J. Plant Sci. Ecol.* 1, 107–112.

41. Kim, H. Y., Lee, S. G., Oh, T. J., Lim, S. R., Kim, S. H., Lee, H. J., et al. (2015). Antiproliferative and apoptotic activity of *Chamaecyparis obtuse* leaf extract against the HCT116 human colorectal cancer cell line and investigation of the bioactive compound by gas chromatography-mass spectrometry-based metabolomics. *Molecules* 20, 18066–18082. doi: 10.3390/molecules201018066

42. El-Hallouty, S. M., Fayad, W., Meky, N. H., EL-Menshawi, B. S., Wassel, G. M., and Hasabo, A. A. (2015). *In vitro* anticancer activity of some Egyptian plant extracts against different human cancer cell lines. *Int. J. Pharmtech Res.* 8, 267–272.

43. Ho, K. L., Chung, W. E., Choong, K. E., Cheah, Y. L., Phua, E. Y., and Srinivasan, R. (2015). Anti-proliferative activity and preliminary phytochemical screening of *Ipomoea quamoclit* leaf extracts. *Res. J. Med. Plant* 9, 127–134. doi: 10.3923/rjmp.2015.127.134

44. Jiju, V. (2015). Evaluation of in vitro anticancer activity of hydroalcoholic extract *of Justicia tranquibariensis. Int. J. Res. Pharm. Biosci.* 2, 10–13.

45. Chaudhary, S., Chandrashekar, K. S., Pai, K. S. R., Setty, M. M., Devkar, R. A., Reddy, N. D., et al. (2015). Evaluation of antioxidant and anticancer activity of extract and fractions of *Nardostachys jatamansi* DC in breast carcinoma. *BMC Complement. Altern. Med.* 15, 1–13. doi: 10.1186/s12906-015-0563-1

14. Anemia: Alive and Kicking, assassinating millions.

Ms. Sakshi Sharma

Nursing Officer (ICN), All India Institute of Medical Sciences Gorakhpur, Uttar Pradesh, 273008, **India**

Abstract: A worldwide public health challenge, iron deficiency anemia affects 2 billion people across all continents, 41.8% of expectant mothers. Anemia leads to high incidences of maternal and infant mortality and morbidity. Anemia is a health problem resulting in a lower-than-normal hemoglobin concentration in the blood or red blood cells in the body. Hemoglobin's role is to transport oxygen. The capacity of the blood to carry oxygen is diminished when hemoglobin levels are low, which is insufficient to support a person's physiological demands. Nutritional deficiencies are one of the most prevalent conditions, and iron is the most deficient micronutrient, mainly among pregnant women. Iron deficiency anemia contributes to detrimental consequences on maternal and infant health. It is a hallmark of both poor nutrition and health. The spectrum of symptoms varies according to the magnitude but generally involves tiredness, weakness, irritability, dizziness, sleepiness, and hair loss. Although taking supplements, including iron and folic acid, is important, the main reason why anemia is still so common is primarily due to poor adherence. One of the major causes of the persistently high prevalence of anemia among pregnant women is a lack of adherence to oral IFA supplementation. Anemic women lack awareness and explanation of the benefits and risks of their medications to ensure proper use. Additionally, the COVID-19 pandemic significantly reduced antenatal care.

Keywords: Hemoglobin, iron deficiency, micronutrient deficiency, nutritional anemia, pregnancy

Introduction

Anemia is defined as the number of red blood cells (RBCs) or hemoglobin in the blood is less than normal.[1] As per the World Health Organization (WHO), anemia is blood hemoglobin of less than 11 gm/dl or a hematocrit of less than 37% in pregnant women.[2] Iron is an essential nutrient required for hemoglobin synthesis which carries oxygen. Less hemoglobin results in a reduction in the oxygen-carrying capacity of the blood, which is insufficient to meet an individual's physiological needs. Mild anemia is defined as a hemoglobin of less than 11 g/dl, moderate as 7–9.9 g/dl, and severe as less than seven g/dl.[3] **Symptoms of Anemia:** Anemia symptoms range from fatigue, irritability, weakness, orthostatic dizziness, headaches, sleepiness, drowsiness, hair loss, exhaustion, exercise-associated dyspnea, and reduced physical and mental capacities, depending on the severity.[4]

Global perspective of anemia: Anemia is a global health concern affecting about 2 billion humans, accounting for 41.8% of pregnant women.[5] South Asia and Africa bear approximately 80% of the anemia load.[6] According to WHO, anemia is responsible for 58% of the cases comprising more than 500 million women in developing nations.[7] As per the National Family Health Survey (NFHS-5) 2019-2021, pregnant women 15-49 years account for 45.7% within the urban areas and 54.3% of the rural sector in India. **The most common anemia in pregnancy:** Low or middle-income countries (LMICs) are at a high risk of anemia during pregnancy because of many factors.[5] Nutritional deficiencies are among the most common causes of anemia. Iron is the most deficient micronutrient, mainly among pregnant women, especially in the South-East Asian region.[8] **Causes of iron deficiency:** Iron deficiency usually occurs after a prolonged negative iron balance resulting from insufficient nutritional intake, decreased absorption from the diet, and accelerated necessities during growth.[9] Additionally, chronic blood loss caused by intestinal parasites, such as hookworm infestation, causes anemia by causing loss of appetite, accelerated movement of food through the digestive tract, increased competition for nutrients, and damaged intestinal walls that hinder the absorption of vitamins, including iron, vitamin B12, and folic acid.[10] The condition worsens as the adolescent period approaches because the menstrual blood flow occurs consonant with the rapid growth and expansion of RBCs. The increased iron needs caused by pregnancy result from tissue growth, blood formation, energy needs, teenage pregnancies, iron loss from postpartum hemorrhage, and repeated pregnancy within two years.[11] **Physiological changes during pregnancy :** Pregnancy is a physiological condition. There is usually no side effect on the overall health of pregnant women. Pregnancy results in hematological and hemodynamic modifications. Although the plasma is disproportionately more significant, the total blood volume increases by approximately 50%, resulting in a fall in hemoglobin caused by dilution of the blood; these changes are beneficial to fight the risk of hemorrhage during delivery.[3] The plasma volume increase is much less in iron-deficient women than in those with optimum iron reserves.[10] **Morbidity & and mortality related to anemia:** Anemia is a major cause of death that can be avoided.[7] Depending on the severity, anemia is directly responsible for 20% and indirectly for 50% of maternal deaths worldwide.[5,11] According to existing research, each 1g/dl increase in hemoglobin in anemic pregnant women reduces the risk of maternal death by 20%, making gestational anemia control and prevention a critical approach to preventing maternal deaths.[7] **Consequences of anemia *On pregnant women:*** Iron Deficiency Anemia (IDA) contributes to detrimental effects on maternal health. It limits working capability and has additional social and economic consequences for the individual and their family.[2] Decreased resistance to infection, prolonged hospitalizations, increased risk for low maternal weight gain, placenta previa, premature rupture of membrane, preterm labor, depleting blood reserves during delivery and the risk of blood transfusions, cardiovascular stress, decreased milk production, and depleted maternal iron stores in the postpartum period.[11] It causes postpartum hemorrhage (PPH) resulting in approximately 500 ml of blood loss.[8,12,13] According to the WHO, PPH is moderate when the blood loss ranges from 500 to 1000 ml and severe when the loss exceeds 1000 ml. Moderate blood loss is not clinically significant except for anemic pregnant women.[14] ***On fetus:***It is a hallmark of both poor nutrition and poor health. It also has an impact on worldwide nutritional issues such as stunting and wasting. This leads to small for gestational age, low birth weight, preterm birth, detrimental impact on placental development, lower iron stores in neonates, and reduced school performance due to poor cognitive and physical development.[2,8,13] Severe anemia increases the risk of neonatal mortality, cardiovascular diseases, and childhood obesity due to a lack of energy to exercise. This makes them incredibly susceptible to various neurodevelopmental, and behavioral abnormalities and psychiatric disorders in later life.[15] **Iron from dietary sources:** A diet high in iron, such as meat, cereals, and green leafy vegetables, is critical for IDA prevention. In a 2000-calorie diet, about 10 mg of elemental iron is present (compared to 65 mg in one 325 mg iron sulfate tablet). An Indian diet can vary from 0.8-4.5 mg/day concerning the type of food consumed, nevertheless, it is estimated that typical absorption from a Western style of eating is 1-5 mg/day.[5]

Physiological iron requirements are highest during pregnancy, and the amount of iron absorbed from a typical diet is insufficient to meet these demands.[12] The nutritional status is further affected by economic and environmental conditions. Poor nutritional knowledge affects the quality of diet. This may be worsened by a lack of urge for food during pregnancy. Additionally, frequent and short pregnancy cycles lead women to move from one pregnancy to the next without replenishing their iron reserves.[16]

Iron requirements during pregnancy

The fetus is at the top of the iron consumption hierarchy.[7] For an average weight of 55 kg, the physiological iron requirement in pregnant women is roughly 1000-1200 mg. The fetus necessitates 270 mg for growth, the placenta requires 90 mg for development, the RBC takes around 500 mg for mass expansion, and the mother will need 250 mg for blood loss at the time of delivery.[17] Iron absorption rises during pregnancy, and roughly 150 mg is conserved owing to amenorrhea, so a healthy pregnant woman can get half of her requirement from a balanced diet. This results in a net need of about 500 mg, which a healthy woman's sufficient iron reserves can meet.[18] To meet the iron demands of pregnancy, women must have more than 500 mg of iron in their stores at the start.[10] Only 20% of reproductive-aged women are thought to have this buffer, and 40% of women, especially in developing countries, initiate pregnancy with no iron reserves.[7] As a result, pregnancy exacerbates anemia and increases the risk of morbidity.[18] To maintain an adequate iron amount and sustain the development of the fetus, women must obtain 1 g of iron during pregnancy.[12] Iron requirements, on the other hand, are not consistent across the three trimesters of pregnancy. Because menstruation stops during the first trimester, the needs (approximated as 0.8 mg/day) are less than before pregnancy. As the fetus grows, maternal erythrocyte mass rises, fetal and placental growth intensifies, and physiologic iron requirements increase to 3-7.5 mg/day in the late pregnancy.[7,17]

Need for oral iron and folic acid supplementation (IFAS) during pregnancy: Supplementing with oral IFAS can also raise the mean blood hemoglobin by 10.2 g/l in pregnant women and 8.6 g/l in non-pregnant women, eradicating nearly half of the anemia in women.[19] Women should take oral IFAS every day while pregnant to avoid various complications due to iron deficiency.[8,12] Non-anemic women should take one oral IFAS containing 60 mg elemental iron and 0.5 mg folic acid daily, whereas anemic women should take two. For optimal absorption, it should be eaten on an empty stomach or 1 hour after mealtime, especially with vitamin C-rich items like orange juice or guava.[7] **The COVID-19 pandemic's impact**: Food availability has changed worldwide because of the pandemic, with varying severity depending on the country's economic situation. In developing nations, this has more serious consequences, such as delays in food delivery, reduced food amount and quality, restrictions on food availability, and loss of revenue to buy food. This has serious immediate health consequences, and the pandemic has resulted in widespread nutritional deficits that will have a long-term deleterious impact on life.[20] Nutrient and caloric intakes were significantly reduced ($P<0.05$).[21] Lockdown harms nutrition and physical activity behavior which might have increased the risk of malnutrition.[22] The pandemic post-ponded the various non-important health services to forestall transmission within clinics, which led to significant reductions in the obtention of antenatal care. An internet survey in the US discovered almost one-third reported increased stress levels, with changes to prenatal appointments as a serious reason for this elevation, and an 18% reduction in ANC was reported. [23,24] 20 million newborns were predicted to be born in India between March and December 2020. The big challenge was to shield pregnant women from being infected while ensuring adequate ANC.[25]

Role of health care provider: Less than a third (28%) received nutrition counseling during their prior prenatal visits. The nutritional advice was only given to women who had poor overall health conditions. This could reduce the projected advancements in nutritional knowledge, which is crucial in the early weeks of development to have a greater effect on health outcomes.[13]

References

1. World Health Organization: WHO. (2019e, November 12). *Anaemia.* Retrieved November 6, 2023, from https://www.who.int/health-topics/anaemia#tab=tab_1

2. Goonewardene, I. M. R., & Senadheera, D. I. (2018). Randomized control trial comparing effectiveness of weekly versus daily antenatal oral iron supplementation in preventing anemia during pregnancy. *Journal of Obstetrics and Gynaecology Research, 44*(3), 417-424.

3. Salhan, S., Tripathi, V., Singh, R., & Gaikwad, H. S. (2012). Evaluation of hematological parameters in partial exchange and packed cell transfusion in treatment of severe anemia in pregnancy. *Anemia, 2012.*

4. Longo, D. L., & Camaschella, C. (2015). Iron-deficiency anemia. *N Engl J Med, 372*(19), 1832-43.

5. Tandon, R., Jain, A., & Malhotra, P. (2018). Management of iron deficiency anemia in pregnancy in India. *Indian Journal of Hematology and Blood Transfusion, 34*, 204-215.

6. Shet, A. S., Zwarenstein, M., Mascarenhas, M., Risbud, A., Atkins, S., Klar, N., & Galanti, M. R. (2015). The Karnataka Anemia Project 2—design and evaluation of a community-based parental intervention to improve childhood anemia cure rates: study protocol for a cluster randomized controlled trial. *Trials, 16*(1), 1-10.

7. Shivalli, S., Srivastava, R. K., & Singh, G. P. (2015). Trials of improved practices (TIPs) to enhance the dietary and iron-folate intake during pregnancy-a quasi experimental study among rural pregnant women of Varanasi, India. *PLoS One, 10*(9), e0137735.

8. Gebremariam, A. D., Tiruneh, S. A., Abate, B. A., Engidaw, M. T., & Asnakew, D. T. (2019). Adherence to iron with folic acid supplementation and its associated factors among pregnant women attending antenatal care follow up at Debre Tabor

General Hospital, Ethiopia, 2017. *PloS one, 14*(1), e0210086.

9. Miller, J. L. (2013). Iron deficiency anemia: a common and curable disease. *Cold Spring Harbor perspectives in medicine, 3*(7), a011866.

10. Ghodeif, A. O., & Jain, H. (2019). Hookworm.

11. Kendre, G. V., & Shinde, R. D. Tele-follow up of anemic ANC Mother; subjective and objective correlation of anemia and potential improvement: Case report study.

12. Jafarbegloo, E., & Tehrani, T. D. (2015). Gastrointestinal complications of ferrous sulfate in pregnant women: A randomized double-blind placebo-controlled trial. *Iranian Red Crescent Medical Journal, 17*(8).

13. Riang'a, R. M., Nangulu, A. K., & Broerse, J. E. (2020). Implementation fidelity of nutritional counselling, iron and folic acid supplementation guidelines and associated challenges in rural Uasin Gishu County Kenya. *BMC nutrition, 6*, 1-16.

14. Diaz, V., Abalos, E., & Carroli, G. (2018). Methods for blood loss estimation after vaginal birth. *Cochrane Database of Systematic Reviews*, (9).

15. Saha, J., Mazumder, S., & Samanta, A. (2018). Does effective counseling play an important role in controlling iron deficiency anemia among pregnant women. *National Journal of Physiology,*

Pharmacy and Pharmacology, 8(6), 840-847.

16. Lartey, A. (2008). Maternal and child nutrition in Sub-Saharan Africa: challenges and interventions. *Proceedings of the Nutrition Society, 67*(1), 105-108.

17. Garzon, S., Cacciato, P. M., Certelli, C., Salvaggio, C., Magliarditi, M., & Rizzo, G. (2020). Iron deficiency anemia in pregnancy: Novel approaches for an old problem. *Oman Medical Journal, 35*(5), e166.

18. Godara, S., Hooda, R., Nanda, S., & Mann, S. (2013). To study compliance of antenatal women in relation to iron supplementation in routine ante-natal clinic at a tertiary health care centre. *Journal of Drug Delivery and Therapeutics, 3*(3), 71-75.

19. Kamau, M. W., Kimani, S. T., Mirie, W., & Mugoya, I. K. (2020). Effect of a community-based approach of iron and folic acid supplementation on compliance by pregnant women in Kiambu County, Kenya: a quasi-experimental study. *PloS one, 15*(1), e0227351.

20. Rodriguez-Leyva, D., & Pierce, G. N. (2021). The Impact of Nutrition on the COVID-19 Pandemic and the Impact of the COVID-19 Pandemic on Nutrition. *Nutrients, 13*(6), 1752.

21. Tulchin-Francis, K., Stevens Jr, W., Gu, X., Zhang, T., Roberts, H., Keller, J& VanPelt, J. (2021). The impact of the coronavirus

disease 2019 pandemic on physical activity in US children. *Journal of Sport and Health Science, 10*(3), 323-332.

22. Visser, M., Schaap, L. A., & Wijnhoven, H. A. (2020). Self-reported impact of the COVID-19 pandemic on nutrition and physical activity behaviour in Dutch older adults living independently. *Nutrients, 12*(12), 3708.

23. Kotlar, B., Gerson, E., Petrillo, S., Langer, A., & Tiemeier, H. (2021). The impact of the COVID-19 pandemic on maternal and perinatal health: a scoping review. *Reproductive health, 18*, 1-39.

24. Roberton, T., Carter, E. D., Chou, V. B., Stegmuller, A. R., Jackson, B. D., Tam, Y., ... & Walker, N. (2020). Early estimates of the indirect effects of the COVID-19 pandemic on maternal and child mortality in low-income and middle-income countries: a modelling study. *The Lancet global health, 8*(7), e901-e908.

25. Zacharias, P., Joseph, M., Jacob, A., Viji, C. L. M., Josy, N., & Lydia, S. (2021). Barriers to antenatal care during COVID-19 pandemic: A hospital-based retrospective cross-sectional study in rural south Karnataka. *Journal of OBGYN, 8*(1), 27-32.

15. Resilience and Technology: Online Learning and Digital Resilience.

Dr. Shikha

Assistant Professor, School of Education, K.R Mangalam University, Gurugram, Haryana, **India**

ABSTRACT: Resilience refers to the ability of an individual, a community, or a system to withstand, adapt to, and recover from adverse situations, challenges, or stressors. It is the capacity to bounce back or "spring back" to a state of normal functioning after facing difficult or traumatic events. The rapid expansion of online learning in recent years has raised important questions about its impact on learners' digital resilience the ability to adapt, persist, and thrive in the digital realm. This study explores the relationship between online learning and digital resilience, investigating the ways in which online education shapes learners' capacity to navigate digital challenges and opportunities. They exhibited the capacity to adapt to various digital tools and to take control of their learning schedules. However, they also experienced digital stress and fatigue, stemming from constant screen exposure and the management of multiple digital resources. Several factors were identified as contributors to digital resilience, including effective digital literacy training, robust social support networks, and regular feedback and assessment mechanisms. These elements played a vital role in fostering resilience and motivation among learners. Conversely, digital overload, limited technical support, and feelings of isolation hindered digital resilience. The implications of these findings highlight the importance of recognizing digital resilience as an integral component of effective online education. Educators and institutions can leverage these insights to create supportive online learning environments that enhance digital resilience and provide targeted interventions to mitigate challenges. This research calls for further exploration of the long-term effects of digital resilience in online learning and the development of strategies to promote resilience in the digital age. This study contributes to our understanding of the complex interplay between online learning and digital resilience and offers valuable insights for educators, policymakers, and researchers seeking to optimize the online learning experience.

Keywords: Resilience, Technology, Online Learning, Digital Resilience

Introduction to Resilience:

Resilience is a dynamic and multifaceted concept that has gained increasing recognition and importance in various fields, including psychology, education, public health, and disaster management. At its core, resilience refers to the ability of individuals, communities, or systems to withstand, adapt to, and recover from adverse situations, challenges, or stressors. It is the capacity to bounce back or "spring back" to a state of normal functioning after facing difficult or traumatic events. However, resilience is not merely about enduring hardships; it also encompasses growth and even thriving in the face of adversity. Resilience is a fundamental aspect of human nature, and it plays a crucial role in how we cope with life's ups and downs. It is the inner strength and resourcefulness that enable individuals to face adversity and emerge stronger. Whether dealing with personal setbacks, health issues, or professional challenges, resilience empowers individuals to weather the storm and emerge on the other side with increased adaptability, emotional strength, and the ability to cope effectively. Key characteristics of resilience include adaptability, emotional strength, coping skills, problem-solving ability, social support, and optimism. Resilient individuals can adjust to changing circumstances, regulate their emotions, and manage stress effectively. They have a repertoire of coping strategies and problem-solving skills at their disposal. Additionally, social support networks, including friends, family, and communities, can significantly enhance an individual's resilience by providing emotional, psychological, and practical support. Resilience is not a fixed trait but rather a dynamic quality that can be developed and strengthened over time. It is a valuable attribute in personal development, as it helps individuals better navigate life's challenges and recover from setbacks. Moreover, resilience is of great importance in understanding and promoting the well-being of individuals and communities facing adversity. The landscape of education is undergoing a rapid and transformative shift, with online learning at the forefront of this evolution. As digital technologies continue to permeate every facet of our lives, the adoption of online learning platforms has gained unprecedented momentum. While this transition offers unparalleled accessibility and flexibility, it has also introduced a new set of challenges for learners. In this context, the concept of digital resilience has emerged as an increasingly vital facet of education, marking the capacity to adapt, persist, and thrive in the digital realm.

Background of Online Learning and Digital Resilience:

Online learning, also known as e-learning or distance education, has witnessed a remarkable surge in recent years. The proliferation of digital resources, the convenience of remote access to educational content, and the opportunities for self-paced learning have made online education a compelling choice for students of all ages and backgrounds. The global COVID-19 pandemic further accelerated this transition, making online learning a necessity rather than a choice for millions of learners worldwide. However, this digital transformation in education is not without its challenges. Online learners encounter a diverse set of hurdles that are distinct from traditional face-to-face education. They must navigate virtual classrooms, manage an array of digital tools, self-regulate their learning, and overcome digital distractions. Consequently, digital resilience—the ability to adapt, persist, and thrive in the digital learning environment—has become an integral component of educational success.

Research Gap and Significance of the Study:

While online learning has become increasingly prevalent, research on the concept of digital resilience is still in its nascent stages. The digital age poses unique demands on learners, and it is essential to comprehend how students cope with these challenges, adapt to the digital landscape, and harness the opportunities it affords. The dearth of empirical research in this domain presents a conspicuous research gap that this study seeks to address. The significance of this research is underscored by the transformative impact of online learning on contemporary education. In an era marked by an ever-increasing reliance on digital technologies, understanding how students develop and utilize digital resilience is not only crucial for their immediate academic success but also for their lifelong learning journeys. The skills acquired and the resilience nurtured in online learning environments hold far-reaching implications for future career prospects, personal development, and adaptability to ever-evolving digital technologies.

Research Objectives: The primary objectives of this study are as follows:

1. To investigate the impact of online learning on students' digital resilience.

2. To identify the factors those contributes to or hinder the development of digital resilience in the context of online education.

3. To explore strategies and interventions that can enhance digital resilience among online learners.

By addressing these objectives, this research aims to provide valuable insights into the interplay between digital education and academic resilience, offering a foundation for evidence-based decisions, interventions, and practices in the ever-evolving landscape of education. It is expected that the outcomes of this study will not only benefit educators, institutions, and policymakers but also contribute to a broader understanding of how learners can thrive in the digital age.

Literature Review: Define Academic Resilience and Digital Resilience: Academic Resilience: Academic resilience is a construct that has been extensively studied in the fields of psychology and education. It refers to the ability of individuals to effectively cope with and overcome adversity in academic settings, enabling them to achieve success despite facing various challenges. Academic resilience encompasses psychological, emotional, and cognitive aspects, such as adaptability, motivation, self-regulation, and the capacity to persist in the face of setbacks. **Digital Resilience:** In recent years, the concept of digital resilience has emerged as an extension of academic resilience in the context of online learning and the digital age. Digital resilience pertains to an individual's ability to adapt, persist, and thrive in digital environments. It encompasses skills like digital literacy, the capacity to navigate online tools, self-regulation in the digital space, and emotional well-being in the face of digital challenges. **2. Discuss the Rise of**

Online Learning and Its Challenges: The Rise of Online Learning: The rapid expansion of online learning, often referred to as e-learning or distance education, is a hallmark of the digital era. It has gained immense popularity due to its accessibility, convenience, and flexibility. Advances in technology, the widespread availability of the internet, and evolving pedagogical approaches have propelled online learning into the mainstream of education. This transition has only accelerated with the global COVID-19 pandemic, which necessitated remote and online education solutions to ensure continuity of learning. **Challenges of Online Learning:** Despite its many advantages, online learning presents a host of challenges for learners. Some of these challenges include: **Self-regulation and motivation:** Online learners must independently manage their schedules, maintain motivation, and adhere to self-imposed deadlines. **Digital literacy:** Navigating online learning platforms, understanding digital tools, and troubleshooting technical issues can be daunting for some students. **Isolation and disconnection:** Online learners may experience feelings of isolation and disconnection from peers and instructors due to the absence of face-to-face interactions. **Digital distractions:** The digital environment offers numerous distractions, from social media to other online activities, which can impede effective learning. **Explore the Existing Research on the Relationship Between Online Learning and Resilience:** The existing body of research examining the relationship between online learning and resilience is in its nascent stages, but several key themes have emerged:

- **Adaptability and self-regulation:** Some research suggests that online learning environments require greater adaptability and self-regulation skills, which can foster resilience in students.

- **Digital literacy and challenges:** Studies have highlighted the significance of digital literacy and the challenges students face in mastering digital tools and platforms.

- **Social support:** Online learners who have access to strong social support networks, including virtual communities and peer interactions, often report higher levels of resilience.

- **Motivation and persistence:** Research has explored the factors that influence online learners' motivation and persistence in the face of digital challenges.

By examining this existing research, this study builds upon the insights gained so far and seeks to contribute to a deeper understanding of the complex interplay between online learning and resilience, with a specific focus on digital resilience.

Methodology:

Research Methods: This study employs a mixed-methods research approach, incorporating both quantitative and qualitative methods to provide a comprehensive understanding of the impact of online learning on digital resilience.

Quantitative Methods: A structured online survey will be administered to a diverse sample of online learners. The survey will include standardized measures and Likert-scale questions to assess digital resilience, self-regulation, adaptability, motivation, and other relevant variables. The quantitative data will be analyzed using statistical software to identify patterns, relationships, and statistical significance in the collected data. **Qualitative Methods:** In-depth semi-structured interviews will be conducted with a subset of survey respondents. These interviews will delve into participants' experiences, challenges, and strategies related to online learning and digital resilience. Qualitative data will be transcribed and analyzed thematically to capture rich narratives and provide context to the quantitative findings. **Study Participants:**The study will target a diverse and representative sample of online learners. Participants will be drawn from various educational levels, including K-12, higher education, and lifelong learning. The sample will include individuals of different ages, backgrounds, and digital literacy levels to ensure a broad perspective on digital resilience in online learning environments.

Data Collection Procedures: Quantitative Data Collection: For the quantitative phase of the study, an online survey will be distributed to potential participants. The survey will be hosted on a secure online platform and will include informed consent information. Participation will be voluntary, and respondents will have the option to remain anonymous. The survey will collect demographic information, academic backgrounds, and responses to items related to digital resilience, adaptability, motivation, self-regulation, and challenges faced in online learning. **Qualitative Data Collection:** A subset of survey respondents will be invited to participate in semi-structured interviews. These interviews will be conducted via video conferencing or voice calls and will be audio-recorded with participants' consent. The interviews will explore their experiences with online learning, the development of digital resilience, and strategies they employ to overcome challenges. Interviews will be transcribed verbatim for thematic analysis. The combination of quantitative survey data and qualitative interviews will provide a comprehensive understanding of the relationship between online learning and digital resilience. The integration of both methods will allow for a deeper exploration of the factors influencing digital resilience while also capturing the nuanced experiences and narratives of online learners. This approach aims to provide a robust foundation for the study's objectives and the development of evidence-based recommendations for educators, institutions, and policymakers. In summary, digital resilience is influenced by a combination of factors, including digital literacy, self-regulation, adaptability, motivation, social support, feedback and assessment, effective course design, technical support, and stress management. These factors collectively empower individuals to adapt, persist, and thrive in the digital environment, ensuring they can effectively navigate digital challenges and opportunities. **Challenges to Digital Resilience:** Digital resilience can be hindered by various factors that online learners may encounter. These challenges include:

1. **Digital Overload:** Online learners often juggle multiple digital platforms, tools, and communication channels. Managing this digital complexity can lead to digital overload, which can be mentally and emotionally draining. Experiencing digital overload can hinder learners' capacity to adapt and remain resilient in the face of challenges.

2. **Technical Support Issues:** While access to technical support is a factor contributing to digital resilience, inadequate or ineffective technical support can be a major challenge. Technical issues that go unresolved or require prolonged assistance can lead to frustration, reducing a learner's resilience in the digital environment.

3. **Feelings of Isolation:** Online learning can sometimes be isolating. Learners may miss the in-person interactions and social connections that traditional classrooms offer. Feelings of isolation can lead to a sense of detachment from the learning experience, making it difficult for individuals to stay motivated and resilient.

Implications and Applications: Interpreting the findings in the context of the literature and theoretical framework provides valuable insights for educators, institutions, and policymakers. Here are some key implications and applications:

1. **Digital Resilience Training:** Educators can incorporate digital resilience training into their courses. This may involve teaching students strategies to manage digital overload, providing guidance on accessing technical support, and fostering a sense of belonging in the digital classroom.

2. **Enhanced Technical Support:** Institutions can invest in robust technical support services to address technical issues promptly. Quick and effective support can reduce digital stress and enhance learners' digital resilience.

3. **Social Interaction and Community Building:** To combat feelings of isolation, institutions can create opportunities for social interaction and community building in the digital learning environment. This can include online discussion forums, group projects, and virtual events that encourage peer engagement.

4. **Balanced Course Design:** Educators should strive for balanced course design that considers the needs of learners and avoids overwhelming them with excessive digital content. Clear instructions, well-structured modules, and the integration of accessible resources can help reduce digital stress and foster digital resilience.

5. **Awareness and Self-Care:** Educators can also promote awareness of digital stress management strategies and self-care. Encouraging learners to establish digital boundaries and prioritize well-being can enhance their digital resilience.

6. **Policymaker Considerations:** Policymakers can consider the role of digital resilience in educational policy. Ensuring that digital literacy and support services are prioritized can contribute to a more resilient digital learning environment.

7. **Research and Continuous Improvement:** Continuous research and assessment of digital resilience are crucial. As technology and online learning evolve, ongoing research can inform the development of new strategies and interventions to enhance digital resilience.

Conclusion:

In this study, we have explored the concept of digital resilience in the context of online learning. Digital resilience, the ability to adapt, persist, and thrive in the digital environment, is increasingly vital in our ever-connected world. The research has identified several factors contributing to digital resilience, including digital literacy, self-regulation, adaptability, motivation, social support networks, feedback mechanisms, effective course design, and access to technical support. These factors collectively empower individuals to navigate digital challenges effectively. However, online learners face challenges that can hinder their digital resilience. Digital overload, technical support issues, and feelings of isolation are among the key challenges. These challenges can cause stress and reduce learners' capacity to adapt and thrive in the digital realm.

Implications of the Research:

The implications of this research are significant and far-reaching. This study provides valuable insights into how educators, institutions, and policymakers can support and enhance digital resilience in online learners. By recognizing the factors contributing to digital resilience and addressing the challenges that hinder it, educational stakeholders can create a more conducive online learning environment. The practical implications include:

1. Incorporating digital resilience training into educational programs.

2. Strengthening technical support services for online learners.

3. Fostering social interaction and community building in the digital classroom.

4. Designing balanced courses that reduce digital stress.

5. Promoting awareness of digital stress management and self-care.

6. Advocating for policies that prioritize digital literacy and support services.

These implications underscore the importance of building resilient online learning environments that empower individuals to adapt and succeed in the digital age. **Closing Statement:** As we conclude this research, it is clear that digital resilience is a dynamic and essential quality for learners navigating the digital landscape. By addressing the factors contributing to digital resilience and mitigating the challenges that hinder it, we can create a more inclusive, supportive, and resilient online learning environment. This research contributes to a growing body of knowledge on digital resilience and offers practical guidance for those involved in education. It is our hope that this work will inspire further research, policy development, and educational practices that enhance digital resilience and foster a more resilient and adaptive digital society.

Refrences

1. Masten, A. S. (2001). "Ordinary magic: Resilience processes in development." American Psychologist, 56(3), 227-238.

2. Bonanno, G. A., & Diminich, E. D. (2013). "Annual research review: Positive adjustment to adversity—Trajectories of minimal–impact resilience and emergent resilience." Journal of Child Psychology and Psychiatry, 54(4), 378-401.

3. Selwyn, N. (2011). "Schools and schooling in the digital age: A critical analysis." Routledge.

4. Al Lily, A. E., Ismail, A. F., & Abunasser, F. M. (2013). "Distance education as a cultural challenge: A Saudi Arabian perspective." Educational Technology Research and Development, 61(3), 483-503.

5. Hodges, C., Moore, S., Lockee, B., Trust, T., & Bond, A. (2020). "The difference between emergency remote teaching and online learning." Educause Review, 27.

6. Rutter, M. (1987). "Psychosocial resilience and protective mechanisms." American Journal of Orthopsychiatry, 57(3), 316-331.

7. Ragnedda, M., & Ruiu, M. L. (2018). "The third digital divide: A Weberian approach to digital inequalities." Routledge.

16. Semiotic Analysis of Language as Propaganda in 'Animal Farm': Insights from Umberto Eco.

Dr. M. Nagalakshmi

Professor & Research Supervisor, Department of English, VELS Institute of Science, Technology and Advanced Studies, Pallavaram, Chennai, **India**

Abstract: This study delves into the intricate relationship between language and propaganda in George Orwell's allegorical novella, "Animal Farm." It employs a semiotic analysis using the theoretical insights of Umberto Eco to examine how language is utilised as a tool of propaganda by the ruling pigs in the story. Through this lens, the research explores the manipulation of signs, symbols, and discourse in the context of an evolving totalitarian regime. Various propaganda techniques are unveiled, and the semiotic interpretation of key symbols and signs is conducted to elucidate their deeper meanings and the psychological impact on the characters. Additionally, the study underscores how language shapes perception, reality, and the power dynamics in disseminating propaganda. This study contributes to both propaganda theory and literary analysis. It deepens our understanding of the propaganda mechanisms within "Animal Farm" and offers insights into the broader implications of language manipulation in totalitarian regimes. Ultimately, this study sheds light on the enduring relevance of "Animal Farm" in contemporary societies and paves the way for future research in propaganda and semiotics.

Keywords: Propaganda, semiotic analysis, totalitarian regime, language manipulation.

Brief overview of George Orwell's "Animal Farm" and its Historical Context. "Animal Farm" is a novella written by George Orwell and first published in 1945. The story is an allegory that serves as a satirical commentary on the events leading up to the Russian Revolution of 1917 and the early years of the Soviet Union. In the novella, Orwell uses a group of farm animals as characters to represent real-life figures and classes of that era. The story begins on Manor Farm, where the animals, led by Old Major, a wise old boar, stage a rebellion against their oppressive human owner, Mr. Jones. The animals succeed in driving Mr. Jones away and renaming the farm "Animal Farm." They establish a set of seven commandments that are meant to guide their new society, which is built on the principles of equality and cooperation. Over time, however, the pigs, who initially played a pivotal role in the rebellion, begin to consolidate power and become increasingly corrupt. They manipulate the commandments, rewriting them to serve their interests, and gradually transform the farm into a totalitarian state ruled by a single party under the leadership of

Napoleon, representing Joseph Stalin. The historical context of the Russian Revolution is crucial to understanding the novella. The revolution in 1917 overthrew the Russian monarchy. It led to the establishment of the Soviet Union under the leadership of the Bolshevik Party, with Vladimir Lenin and later Joseph Stalin at the helm. "Animal Farm" is a critical allegory that reflects the events and betrayals that occurred during this historical period, illustrating the dangers of totalitarianism and the manipulation of language and propaganda. **Introduction to the Role of Language and Propaganda in the Novella.** In "Animal Farm," the manipulation of language and the use of propaganda are central themes that significantly impact the story's development. George Orwell, in his exploration of these themes, offers a poignant commentary on the power dynamics within authoritarian regimes and the manipulation of information to control the masses. Language is not merely a means of communication in the novella; it becomes a tool for those in power to maintain their authority and suppress opposition. One of the pivotal quotes in the novella that underscores the importance of language is the modification of the Seven Commandments. The original commandments, which symbolise the ideals of the animal rebellion, are altered to suit the pigs' changing agenda. As they begin to violate the commandments, they also subtly change their wording. For instance, the original commandment "All animals are equal" is transformed into "All animals are equal, but some animals are more equal than others" (Orwell, 1945, p. 112). This alteration not only highlights the hypocrisy of the ruling class but also demonstrates how language is used to justify inequality and maintain control. Another key element is the clever use of propaganda by the pigs, notably through slogans and songs. The catchy slogan "Four legs good, two legs bad" (Orwell, 1945, p. 56) is repeatedly chanted by the animals to reinforce the notion that humans are the enemy. This slogan simplifies a complex issue into a binary opposition, appealing to the animals' emotions rather than their critical thinking. Similarly, the song "Beasts of England" initially serves as an anthem of hope and rebellion but is later replaced by the pig-authored "Animal Farm" song, which promotes loyalty to the new regime. These instances exemplify the manipulation of language and the power of propaganda to shape the animals' perceptions and allegiances. Language becomes a tool for the pigs to maintain their authority and consolidate power, all the while creating a façade of equality and liberation.

Research Problem

The central research problem of this study is to investigate the manipulation of language for propaganda in George Orwell's "Animal Farm." The novella portrays the gradual distortion of language and the use of propaganda as tools of control by the ruling pigs. Understanding how language and propaganda are manipulated in the text is crucial to revealing the underlying mechanisms of propaganda and totalitarianism.

Research Objectives: The specific objectives of this study are as follows:

1. **To analyse the evolving role of language in "Animal Farm"**: This objective aims to dissect the transformation of language from a medium of revolution to a tool of manipulation and control. It will involve a detailed examination of key linguistic elements and their impact on the characters and the story.

2. **To explore the use of propaganda in the novella**: This objective seeks to investigate the various propaganda techniques employed by the ruling pigs, including slogans, the dissemination of information, and the creation of a cult of personality. The study will shed light on how propaganda shapes the animals' perceptions.

3. **To interpret the implications of linguistic manipulation and propaganda in the novella**: This objective will focus on the broader implications of language and propaganda in "Animal Farm." It will address the consequences of these manipulations on the characters, the society, and the narrative's development.

4. **To apply Umberto Eco's semiotic analysis to uncover hidden meanings and symbols**: Drawing upon the semiotic framework of Umberto Eco, this objective involves the application of semiotics to the novella. It aims to reveal hidden meanings, symbols, and signifiers within the text that contribute

to a deeper understanding of the propaganda mechanisms at play.

5. **To assess the effectiveness of Eco's semiotic approach**: This objective evaluates the utility of Umberto Eco's semiotic framework in analysing the linguistic and propagandistic elements in "Animal Farm." It seeks to determine how semiotics enhances our understanding of the text and its socio-political commentary.

Explanation of Key Concepts in Umberto Eco's Semiotics

Umberto Eco's semiotics, as articulated in his seminal work "A Theory of Semiotics" (1976), encompasses a rich framework for understanding the nature of signs and the construction of meaning. Key concepts within Eco's semiotic theory include signs, signifiers, and signified. **Signs:** A fundamental concept in Eco's semiotics is the sign. A sign is a basic unit of communication, representing something that carries meaning. According to Eco, a sign is a "two-faced entity, composed of a 'signifier' (the form) and a 'signified' (the content)." "Semiotics is concerned with everything that can be taken as a sign." (Eco, 1976, p. 7). In other words, a sign is the combination of a material form (the signifier) and a concept or mental image (the signified). **Signifiers:** Signifiers are the material or physical aspects of signs. They are the concrete representations that are perceived through our senses. For example, in written language, the letters and words themselves are the signifiers. The signifiers provide the physical form through which the sign conveys meaning. "A sign is always the combination of a 'signifier' and a 'signified'." (Eco, 1976, p. 7). **Signified:** The signified is the conceptual or mental content associated with a sign. It represents the idea, concept, or meaning that the sign conveys. In the context of written language, the signified is the mental image or concept evoked by the arrangement of letters and words, as interpreted by the reader. "A sign is a two-faced entity, composed of a 'signifier' (the form) and a 'signified' (the content)." (Eco, 1976, p. 7) This framework provides a valuable tool for analysing how signs are employed in various forms of communication, including literature and art. **Identification of Key Symbols and Signs in the Novella "Animal Farm"** "Animal Farm" is rich in symbols and signs that carry deeper meanings, reflecting the novella's allegorical nature. Several key symbols and signs can be identified in the narrative. The Seven Commandments represent a set of signs that symbolise the original ideals of the animal rebellion. They are prominently displayed on the side of the barn, signifying the principles of equality and animal solidarity. These commandments are crucial symbols that reflect the animals' initial vision of a just society. "All animals are equal." (Orwell, 1945, p. 4) The windmill is a significant symbol in "Animal Farm." It initially represents progress and the animals' efforts to build a better future. However, the windmill is repeatedly destroyed and then reconstructed, becoming a symbol of the ruling pigs' manipulation and propaganda. It exemplifies how symbols can be transformed to convey different meanings. "The windmill was in ruins." (Orwell, 1945, p. 100) The flag, with a hoof and horn as its emblem, serves as a sign of the animal rebellion. It is a symbol of unity and resistance against human oppression. The flag conveys the animals' collective identity and the ideals of their revolution. "The flag was green, Snowball explained, to represent the green fields of England, while the hoof and horn signified the future Republic of the Animals." (Orwell, 1945, p. 6) The altered Seven Commandments reflect the manipulation of signs to suit the pigs' evolving desires. They signify the corruption of the original ideals and the pigs' distortion of the animals' vision. The changing commandments are signs of ideological manipulation and propaganda. "All animals are equal, but some animals are more equal than others." (Orwell, 1945, p. 134) These symbols and signs in "Animal Farm" are integral to the narrative's allegorical nature, representing concepts, ideals, and manipulation. Their identification and interpretation are essential for a comprehensive understanding of the novella. **The significance of applying Umberto Eco's semiotic analysis to the text.** Umberto Eco's semiotic analysis is particularly relevant and valuable for understanding the deeper layers of meaning in "Animal Farm." Eco's semiotics offers a structured framework for deciphering the subtle and often hidden layers of meaning in literary texts, making it a powerful tool for unravelling the intricacies of Orwell's allegorical work. Eco's semiotics is characterised by its emphasis on the study of signs and symbols as carriers of meaning. Eco argued that signs are not mere arbitrary entities but rather components of a complex system of communication. This approach enables the exploration of the various dimensions of a text, including its surface meaning and the underlying symbols and messages. In "Animal Farm," the application of semiotic analysis is highly significant for several reasons: Uncovering Hidden Symbols: Eco's semiotics can reveal the presence of hidden symbols and signifiers within the text, which may not be immediately apparent. For example, the novella's use of the windmill can be interpreted as a multifaceted symbol, representing both the animals' aspirations and the pigs' manipulation. Eco's framework allows us to identify and interpret such symbols. "A sign can refer either to something with the same form (in this case, it is an icon) or to something with a different form (in this case, it is a symbol), but it can also refer to something with a sort of form that makes it similar to other things (in this case it is an index)." (Eco, 1976, p. 23) Deeper Understanding of Propaganda: Semiotic analysis can uncover the layers of meaning embedded in propaganda techniques and slogans used by the pigs. Eco's framework provides tools for dissecting the subtle ways in which language is employed to manipulate and control, allowing readers to grasp the true intentions behind propaganda. "Words are the most powerful drug used by mankind." (Rudyard Kipling, as quoted by Eco, 1976) Interpreting the Animals' Actions: Eco's semiotics aids in the interpretation of the animals' actions, shedding light on their motivations and the underlying messages conveyed through their behaviour. By understanding the semiotic codes of the text, readers can uncover the nuances of character development and plot progression. "A message always has both a form and a content, and the interpretation depends on the cooperation of the two levels of meaning." (Eco, 1976) Enhancing Literary Analysis: Applying semiotic analysis enhances the literary analysis of "Animal Farm" by providing a structured method for exploring the text's complexities. This approach enriches the reader's appreciation of the novella's intricate layers of meaning and socio-political commentary. "Semiotics tells us things we already know in a language we don't understand." (Umberto Eco, as quoted by Eagleton, 1983)

Conclusion

In the course of this study, several key findings and insights have emerged, shedding light on the manipulation of language for propaganda in George Orwell's "Animal Farm" through the lens of Umberto Eco's semiotic analysis. Firstly, the study revealed how the pigs, the ruling class in the animal society, strategically manipulated language, transforming it from a tool of collective empowerment into one of control and manipulation. This manipulation is epitomised by the distortion of the Seven Commandments, where "All animals are equal" was transformed into "All animals are equal, but some animals are more equal than others" (Orwell, 1945, p. 134). Language serves as a malleable tool for altering the perception of reality, and the manipulation of the commandments illustrates this transformation vividly. Secondly, symbols and signs within the novella played a pivotal role in conveying propaganda. Key symbols such as the Seven Commandments, the windmill, and the flag represent not only the animals' ideals but also their corruption and distortion under the pigs' rule. These symbols reflect how language and semiotics are used to control the animals' understanding of their society, as seen in the evolving interpretation of the windmill as a propagandistic spectacle (Orwell, 1945, p. 99). Moreover, Umberto Eco's semiotic framework provided a structured approach to the analysis of these symbols and signs. By focusing on signifiers and signified, the framework unveiled deeper layers of meaning within the text. It allowed for the interpretation of how linguistic manipulation and propaganda function within "Animal Farm". In conclusion, this study has provided a comprehensive analysis of how language and semiotics are employed for propaganda in "Animal Farm." This emphasises how crucial critical analysis is to understanding texts' deeper meanings and how they affect society.

Critical Awareness and Societal Relevance:

The importance of understanding the role of language and semiotics in "Animal Farm" extends beyond the novella itself. It encourages readers to develop a critical awareness of the ways language can be used to shape reality and control narratives. Such awareness is crucial in contemporary society, where misinformation and propaganda are prevalent. By examining the past, we can better navigate the complexities of the present. "No question, now, what had happened to the faces of the pigs? The creatures outside looked from pig to man, and from man to pig, and from pig to man again, but already it was impossible to say which was which." (Orwell, 1945, p. 141) Thus, comprehending the role of language and semiotics in "Animal Farm" is essential to recognise the potential misuse of language and signs for propaganda and manipulation. It underscores the novella's enduring relevance as a warning about the dangers of authoritarianism and deceptive rhetoric, urging us to remain vigilant in our contemporary world.

References:

Orwell, G. (1945). Animal Farm. Secker and Warburg.

Secondary Sources:

Barthes, R. (1977). Image, Music, Text. Hill and Wang.

Eagleton, T. (1983). A Sign of the Times: The Portrayal of Some Distinguished Italians in English Literature. Verso.

Eco, U. (1976). A Theory of Semiotics. Indiana University Press.

Eco, U. (1984). Semiotics and the Philosophy of Language. Indiana University Press.

Figes, O. (1997). A People's Tragedy: The Russian Revolution, 1891-1924. Penguin Books.

Fitzgerald, F. S. (1925). The Great Gatsby. Charles Scribner's Sons.

Foucault, M. (1972). The Archaeology of Knowledge. Pantheon Books.

Hawthorne, N. (1850). The Scarlet Letter. Ticknor, Reed, and Fields.

Hedges, C. (2003). Language as Theme and Art in Animal Farm. In G. Orwell & H. Bloom (Eds.), George Orwell's Animal Farm (pp. 101–121). Chelsea House Publications.

Smith, J. A. (2010). Manipulation of Language for Propaganda in George Orwell's Animal Farm. *Journal of Literary Analysis, 5*(2), 45–59.

17. The state of Andhra Pradesh's Use of New Media in the Digital Age: A Mediating Role.

Dr. T. Yogesh Babu[1] and Dr. P. John Adinarayana[2]

[1]Former Assistant Professor &Head, Department of Fine Arts, EIT, CASS, CBSS, Mainefhi ,Adikeih, Eritrea , North East Africa , Africa, IIP PDF-2023 candidate from ERU- USA& ERC-**India**

[2]Assistant Professor, Department of Fine Arts, Koneru Lakshmaiah University, Guntur, AP, **India**

Abstract: All web-related technologies, including blogs, social networking sites, online social media networking, and other forms of communication technology, are referred to as new media technologies. This study's main goal is to investigate how social media and new media interaction function as mediators in the current digital era in the state of Andhra Pradesh. This study also examines the definition of new media, outlines the features of new media technologies, and examines how new media are applied in the current digital era. In order to achieve the aforementioned goals, the researchers administer a questionnaire survey to demonstrate how new media engagement acts as a moderator between the directives given to newsrooms in the current digital era and the public's perception of social media.

Keywords: Use of new media technologies, Social networking, Digital age, communication technologies, attitude toward social media.

Introduction

The role that new media technologies play in a variety of fields, such as business, media, education, mass communication, and others, has recently attracted increasing interest. Web 2.0 technologies are a common term used to describe new media technologies. In order to communicate with a wider audience, these technologies make use of web technologies such as blogs, wikis, online social networking, and other social media platforms. Old media, including magazines, television, newspapers, radio, and other traditional media forms, have seen a sharp fall in use in recent years, while the internet has quickly supplanted old media. As a result, traditional mass communication methods are rapidly disappearing from existence. Numerous studies, such as those by Lysak et al., Andrus, and Boulos et al., have shown how the telegraph was replaced by the telephone and email, life magazines by television, and so on. These examples demonstrate how current communication methods have supplanted traditional media. By operating concurrently to satisfy the needs of many organizations, new media communication technologies complemented the various kinds of traditional media. Any kind of mass communication can now be found at newsstands and online. Readers of mass communication media are able to post comments on blogs and online social networking sites instantly. Advances in the field of new media technologies enable people to take part in public conversations. Mass communication technology as a whole has changed with the recent developments in the field of technology. The process of communication has taken on a whole new form as people started to relate to new media on a personal level, which was not the case with older media. In order to better understand the role that new media technologies play in helping people create their own identities in the digital era, many scholars have emphasized how widespread the idea of virtual identity has become in recent years. The researchers will be able to study how people create their own identities through communication on social networking sites with the use of this information. The ability of new media technology to engage users in public debates on politics or society is another fascinating feature. Actually, these technologies give users a private area where they can voice their opinions in the virtual public realm created by computers. Users are now able to create unfettered internet content thanks to new media technologies. Initial research on new media technologies mostly concentrated on the constrained consequences of contemporary communication methods. In the media, for example, very few reporters for newspapers will be able to comprehend the importance of blogs; very few producers of television will be able to comprehend why viewers switched from network television to internet television, and so on. An increase in the digitization of knowledge was documented in a recent study on New Media technologies. It was discovered that the digital transformation through the mobile-first strategy was driven by consumers searching for information regardless of the location, time, or convenience. Here are some further intriguing discoveries in this field:

(a) Young people are increasingly consuming information on social media platforms;

(b) People are starting to check information online;

(c) Businesses are promoting their brands, products, and services through these and other social media platforms;

(d) Media professionals are using new media to find new audiences, double-check Information, support news coverage, and check the credibility of the news;

(e) Advertising and marketing channels are using new media technology to create branding for their goods and services; and

(f) Academics are starting to use various new media technologies for teaching purposes.

According to recent studies, the percentage of persons utilizing different forms of new media technology has increased from 15% in 20018 to 68% in 2022. In an effort to comprehend new media technologies' influence on the Digital Age, scholars have been concentrating on their function for a number of years. It has been shown that computer-generated public spheres assist new media technologies in content creation. In the digital age, these tools are thought to be an efficient tool for sustaining engagement. These technologies also offer cost-effectiveness and immediate benefits. One recent study, for instance, showed evidence of users going live on the Facebook app, which enabled them to reach a far greater audience than anticipated. It has been underlined that although there is no method to quantify the engagement of new media, people's involvement, the connectivity between individuals, and considering the degree to which they are willing to participate will aid the researcher in identifying and quantifying the mediating role that engagement plays in this digital age. Organizations across a variety of industries are able to track the part that people play in the activities that are completed on new media platforms such as blogs, wikis, online social networking sites, and other social media platforms with the use of cutting-edge technologies like web analytics. One can track the degree of involvement by counting the likes, shares, views, and comments. Identification of COVID-19 infected individuals with the use of machine learning algorithms to process travel and healthcare data in lieu of the conventional healthcare system This study paper's overall goal is to examine how new media interaction functions as a mediator in the current digital world. The current study is organized as follows: An introduction to the subject is given in Section Introduction; Section Literature Review looks at the body of research on new media as a concept, its features, and its effects on people, organizations, and other domains. The use of new media technologies across disciplines is also covered in this section. The study's methodology is presented in Section Methodology. In Section Data Analysis, the study's conclusions and outcomes are covered. Lastly, the study's conclusions are covered in Section Results and Discussion.

Literature Review: Understanding New Media: The advent of new media has ushered in a transformative era in the way we create, disseminate, and consume information and entertainment. This literature review delves into key concepts and insights that help us better understand the dynamics and implications of new media in the 21st century.

Defining New Media: New media encompasses a broad array of digital technologies and platforms, including social media, mobile apps, streaming services, and interactive websites. Jenkins (2006) defines new media as participatory, interactive, and collaborative, marking a shift from traditional one-way communication to a multidirectional exchange of information. The rise of UGC, facilitated by platforms like YouTube, Instagram, and Wikipedia, has empowered individuals to create and share content. Burgess and Green (2009) highlight the significance of UGC as a democratizing force in media, allowing ordinary people to contribute to the public discourse. Henry Jenkins (2006) introduces the concept of convergence culture, where media industries, technologies, and cultures intersect. This convergence leads to the blurring of lines between content producers and consumers, transforming how we engage with media. Boyd (2014) explores the role of social media in shaping identities and relationships. Platforms like Facebook and Twitter have become central to self-presentation, connection, and identity construction. Greenhow, Robelia, and Hughes (2009) emphasize the importance of digital literacy in the digital age. They argue that education must adapt to teach critical digital literacy skills, enabling individuals to navigate and participate in new media landscapes effectively. Castells (2009) discusses the global impact of new media, highlighting the ability of digital communication to transcend geographic and cultural boundaries. New media has facilitated global activism, information sharing, and cross-cultural dialogue. The psychological impact of new media is a subject of study. Research by Rosen (2013) delves into the effects of digital technologies on attention spans, multitasking, and social relationships, highlighting both advantages and potential drawbacks. The ethical dimensions of new media, especially concerning privacy and data security, have garnered attention. Boyd (2010) explores the complex privacy practices and challenges in the age of social media, emphasizing the need for responsible data handling. Benkler (2006) discusses the implications of participatory culture and the tension between open, collaborative networks and traditional media conglomerates. This raises questions about who controls information and how. Sunstein (2017) examines the impact of new media on political discourse and polarization. He discusses the echo chambers and filter bubbles that can arise in online environments and their influence on public opinion. In conclusion, new media has redefined the ways we communicate, share information, construct identities, and engage with the world. This literature review reflects the diverse aspects of new media, from its participatory nature to its societal, psychological, and ethical dimensions. Understanding new media is critical for navigating the complex and ever-evolving digital landscape in the 21st century.

Proposed hypothesis of the study

H1. Most of the users in the state of AP is aware of new media technologies

H2. The characteristics of new media technologies positively influence the user behaviour

H3. There is a relation between new media technologies and digital age

H4. Almost all the people in this digital age use web technologies for communication

H5. New media technologies enable interactivity among users

H6. New media plays a mediating role in this digital age

Methodology

The primary goal of the current study is to investigate how new media interaction functions as a mediator in the digital age. In addition, the notion of new media is explored, along with the various varieties and attributes of new media technologies. The study looks at new media applications in the digital era as well. Secondary literature as well as primary analysis have been used in the analysis of this study. Based on the secondary literature's analysis and review, the study has generated some possibilities. A questionnaire is used to test the proposed hypotheses. Because a questionnaire is used to collect all necessary data, this study is descriptive in nature. The researcher will assess the hypotheses based on the survey data.

The population from AP region is nothing more than the entire collection of people or things that have certain things in common. There are two categories of population: accessible and target. The people for whom the researcher hopes to generalize the study's findings make up the target group. As a subset of the population, on the other hand, is the other set. All individuals who are under the age of fifty are included in the total population for the purposes of this study. The study obtains the necessary sample from the entire population. The people chosen to take part in the study are included in the sample. These chosen individuals are commonly referred to as study subjects or respondents. There are two categories of sampling techniques: non-probability and probability/random sampling.

In order to achieve the goals of the research, a sample of 200 hundred respondents aged between 16 − 45 is chosen only from AP region. To determine the outcome, main and secondary data are gathered.

Primary data gathered with the use of a structured questionnaire intended to get the necessary information from the subjects. The gathered information will assist the researcher in analyzing the factor of engagement and interaction among users with a special focus on the digital age, reviewing the attitude or behavior of users who use new media technologies to communicate, and examining the mediating role of new media engagement in this digital age. The secondary data needed for the study is gathered by the researcher from journals and research articles. Additionally, the study uses a 3-point Likert scale to assess the developed hypothesis. A preliminary investigation is carried out to assess the validity and reliability of the survey, involving 25 participants. In the end, all the data was compiled and examined to provide an answer.

Data Analysis

The information obtained from the chosen respondents via the questionnaire is presented in this portion of the article. First, the respondents' demographic information is provided, together with information about the duration, context, and other pertinent factors. The study ran from the end of August until the beginning of October 2023. The questionnaire was handover to each of these 250 responders. Since they had to complete out the questionnaire correctly, every participant gave the questionnaire their whole attention. Only 230 of the 250 participants—who included young people and workers in a variety of fields including marketing, media, press, education, and self-employment—received the questionnaire, of which 200 were valid.

	fr	%
MALE	100	50.0
FEMALE	100	50.0
Total	200	100.0

Table 1 Gender

Understanding New Media Technology

About new media technology	Respondents response in %'s
Partially aware of new media technology	30
Un ware of new media technology	15
Fully aware of new media technology	40
Neutral	15
Total	100

Table 2 Concept of new media technology

From Table 2, it can be deduced that of all participants, 40% of them reported that they were fully aware of the term and concept of new media technology, 30% of them reported that they were partially aware of the term and concept of new media technology, 15% of them reported that they were unaware of term and concept of new media technology, and another 15 % of them stayed neutral.

Characteristics of New Media Technologies

Characteristics of New Media Technologies	Respondents response in %'s
Communication	30
Community	10
Creativity	30
Collaboration	30
Total	100

Table 3 Characteristics of New Media Technologies

From Table 3, it can be inferred that out of 200 participants, communication features cater to 30%, convergence and community cater to 10%, the creativity of the users cater to 30%, and collaboration caters to 30%, which will positively affect the behavior of an individual/user in this digital age. This shows that almost all characteristics of new media technology impact the behavior of an individual.

New Media Technologies and Digital Age

New Media Technologies and Digital Age	Respondents response in %'s
Positive relationship	70
Negative relationship	20
Neutral	10
Total	100

Table 4 New Media Technologies and Digital Age

From Table 4, it can be inferred that out of 200 participants 70% of them agreed that there is a positive relationship between new media technology and the digital age, while 20% of them agreed that there is a negative relationship between new media technology and the digital age and 10% stayed neutral. This shows that there certainly exists a relationship between new media and the digital age.

Mediating Role of Engagement and Interactivity

Mediating Role of Engagement and Interactivity	Respondents response in %'s
New Media Engagement	80
New Media Interactivity	15
Neutral	5
Total	100

Table 5 Mediating Role of Engagement and Interactivity

From Table 5, it can be inferred that out of 200 participants 80% of them agreed that new media plays a mediating or engaging role in this digital age to share information, view content, post feedback on blogs, and so on. Meanwhile 15% of them agreed that interactivity is essential in new media technologies, especially in this digital age, and hence interactivity plays a mediating role. Only 05% of the respondents stayed neutral. This shows that both engagement and interactivity play a mediating role between new media and the digital age. New media engagement is essential to share information with each other, develop creative content, and create branding, and so on. Figure 4: Demonstrates the evaluation of the measurements for several parameters. In this digital age, the mediating role of engagement between new media and information was examined in the current study. We were also able to determine whether the respondents were aware of the new media technologies that are quickly gaining traction in this digital age by looking at their demographic information. The findings imply that awareness of new media web-communication technologies is widespread among those living in the digital age. Additionally, the findings show that nearly 90% of respondents believe that the five main features of new media—community, creativity, collaboration, and communication—are what shape people's use of new media in the twenty-first century. Furthermore, as everything is now done online, these qualities are universal and applicable to everyone. The study also discovered a favorable correlation between new media technology and the digital era in AP region. In the digital age, people can use various new media web-communication technologies into their daily lives to facilitate smoother and easier communication. According to the study's findings, new media is important in today's digital world for sharing information, consuming material, leaving comments on blogs, and other activities. The study also discovered that interaction has a mediating role because it is crucial to new media technologies, particularly in this digital era. The digital age and new media are mediated by both interaction and engagement. Interaction with new media is necessary for information sharing, the creation of original content, branding, and other purposes. It is possible to view the current study as only the beginning, with further research to examine the mediating function of additional factors in the digital age. Increasing the number of responders is also advised in order to obtain additional data for this investigation. The dissemination of the message to the intended audience is significantly impacted by user involvement. Similarly, research indicates that the majority of people in the digital age share their activities on numerous social media sites. To put it briefly, user involvement is crucial to the processing of all types of data. Andhra Pradesh People's behavior is changing as a result of these new media technologies, which allow individuals to engage in social interactions, build online communities, and connect online. Therefore, one may argue that audience, user, and individual interaction is essential to new media technology communication.

Conclusion

The mediating role of new media engagement in this digital age in the state of Andhra Pradesh has been proposed and tested in this article. Since interactivity is crucial to new media technologies, particularly in the digital age, it serves as a mediator. The digital age and new media are mediated by both interaction and engagement. Interaction with new media is necessary for information sharing, the creation of original content, branding, and other purposes. The features of new media technologies that impact communication processes in all types of organizations were examined in this study. The study's findings showed that people's behavior is influenced by new media technologies, which make it possible for people to engage in social interactions, build online communities, and interact online in the state of Andhra Pradesh. Therefore, one could argue that audience, user, and individual engagement is essential to new media technology communication. Researchers in the future can use a scientific method to comprehend how new media engagement functions as a mediator in the current digital era. By conducting case studies, they can concentrate on the interactive component and produce results that are much more appropriate and applicable to a variety of fields.

eferences

1. Jenkins, H. (2006). Convergence culture: Where old and new media collide. New York University Press.

2. Boyd, D. (2014). It's complicated: The social lives of networked teens. Yale University Press.

3. Benkler, Y. (2006). The wealth of networks: How social production transforms markets and freedom. Yale University Press.

4. Greenhow, C., Robelia, B., & Hughes, J. E. (2009). Learning, teaching, and scholarship in a digital age: Web 2.0 and classroom research: What path should we take now? Educational Researcher, 38(4), 246-259.

5. Castells, M. (2009). Communication power. Oxford University Press.

6. Sunstein, C. R. (2017). #Republic: Divided democracy in the age of social media. Princeton University Press.

7. Rosen, L. D. (2013). Reclaiming conversation: The power of talk in a digital age. Penguin Books.

8. Burgess, J., & Green, J. (2009). YouTube: Online video and participatory culture. John Wiley & Sons.

9. Boyd, D. (2010). Social network sites as networked publics: Affordances, dynamics, and implications. In A networked self: Identity, community, and culture on social network sites (pp. 39-58). Routledge.

10. Jenkins, H. (2008). Confronting the challenges of participatory culture: Media education for the 21st century. The MIT Press.

11. Boyd, D. (2010). Making sense of privacy and publicity. In A networked self: Identity, community, and culture on social network sites (pp. 169-197). Routledge.

12. Castells, M. (2010). The rise of the network society (Vol. 1). John Wiley & Sons.

13. Green, N. (2019). The internet: An ethnographic approach. Bloomsbury Publishing.

14. Livingstone, S. (2014). The challenge of changing audiences: Or, what is the audience researcher to do in the age of the internet? European Journal of Communication, 19(1), 75-86.

15. Turkle, S. (2015). Reclaiming conversation: The power of talk in a digital age. Penguin.

16. boyd, d., & Ellison, N. B. (2007). Social network sites: Definition, history, and scholarship. Journal of Computer-Mediated Communication, 13(1), 210-230.

17. Lister, M., Dovey, J., Giddings, S., Grant, I., & Kelly, K. (2009). New media: A critical introduction. Routledge.

18. Rheingold, H. (2012). Net smart: How to thrive online. MIT Press.

19. Van Dijck, J. (2013). The culture of connectivity: A critical history of social media. Oxford University Press.

20. Bruns, A., & Burgess, J. (2011). The use of Twitter hashtags in the formation of ad hoc publics. In Proceedings of the 6th European Consortium for Political Research (ECPR) General Conference.

18. A scrutiny study between Shopee and Lazada User Experience and Satisfaction among Filipino Customers.

Rhonnel S. Paculanan[1], Angelo C. Arguson[2], Marvin P. Mabboran[3], Jerie Vale Baustista[1] and Marisol De Guzman[1]

[1]Arellano University, **Philippines**

[2]FEU Tech, Manila, **Philippines**

[3]DepED, SDO Manila, **Philippines**

Abstract: The emergence of e-commerce has had a huge influence on worldwide society, changing how individuals trade and purchase things. The COVID-19 epidemic hastened the switch to online purchases and raised e-commerce sales in the Philippines. Mobile devices are quickly dominating online shopping, and digital businesses must adapt to meet the needs of mobile customers. A mobile e-commerce platform's user experience (UX) must be fluid, simple to use, and well-designed. As a result, the purpose of this study was to assess and compare the UX satisfaction of Filipino customers. The User Experience Questionnaire

Short Version (UEQ), System Usability Scale (SUS), and Net Promoter Score (NPS) were utilized in the study to assess UX satisfaction. The online study of 120 Filipino customers discovered substantial disparities in UX and satisfaction between the Shopee and Lazada mobile applications. The findings of this study add to a better understanding of the importance of UX in mobile e-commerce satisfaction. They can assist in improving the user experience of mobile applications for Filipino customers.

Keywords: Comparative Study, User Experience, User Satisfaction, Customer Loyalty, Shopee, Lazada

Introduction

Today, the global society increasingly relies on e-commerce, which has substantially altered how individuals trade and acquire things. E-commerce has evolved as a significant retail model and online buying has risen in prominence as information technology has advanced (Huang, 2022). According to Keenan (n.d.), the global e-commerce industry is expected to reach $5.55 trillion in 2022, and this figure is expected to rise in the next years, illustrating how lucrative borderless e-commerce is becoming for online retailers. The pandemic of COVID-19 has had a profound impact on worldwide e-commerce patterns. When physical stores shuttered, customers turned to the internet to make purchases. The pandemic, according to experts, expedited the shift to internet purchases by up to five years (Keenan, n.d.). Due to the

COVID-19 pandemic, more Filipinos are staying at home and shopping online, resulting in $17 billion in e-commerce industry sales in the Philippines in 2021, with 73 million monthly active users, and expected to grow at a 17% annual rate to $24 billion in 2025 (Philippines - eCommerce, n.d.-b). Shopee, Lazada, Zalora, and Beauty MNL are the top e-commerce sites in the Philippines. The majority of the working population (ages 25 to 44) actively uses their desktop and mobile devices to access these platforms, and the most popular product categories inside are household care, electronics, fashion, and beauty products (Philippines - eCommerce, n.d.-b). Yatprom (2021) claimed that with the advent of e-commerce, mobile devices are gradually dominating the online retail landscape each year. Since the launch of Amazon's mobile shopping app, mobile devices have accounted for more than 40% of all purchases made on the company's website. Experts expect that this figure will only rise as demand for mobile devices rises, making it critical for digital enterprises to adapt to mobile customers if they want to remain relevant in the digital sphere. Smartphone household penetration in the Philippines rose by 2% between 2020 and 2021, reaching 74.1% (Philippines - eCommerce, n.d.-b). The user experience (UX) of a mobile e-commerce platform must be fluid, easy to use, and well-built since this is critical to the success or failure of any e-commerce venture Perlman (2021). The industry norm is now simple designs with less intricate and confusing material. Many goods on the main page may look OK on a desktop computer, but not on a mobile device since there is little space, and content space is far more precious. Furthermore, crowded pages load more slowly on mobile devices, potentially significantly slowing down the loading time (Yatprom, 2021). Designers may examine the usability of a design to determine engagement, however, due to the expensive expense of usability testing, many organizations make little to no effort to assess the usability of the goods and services they provide (Adobe, 2020). As one of the Philippines' rising business sectors, increasing the user experience and happiness of the top two mobile e-commerce platforms is critical to its growth and development. As a result, this study focuses on analyzing and comparing the user experience and happiness of Filipino customers toward Shopee and Lazada Mobile apps, as well as the importance of user experience to user satisfaction.

Objectives of the Study

This survey assesses the entire user experience and happiness of Filipino customers with the Shopee and Lazada mobile applications. It looks for any significant changes between app users in these two variables. To evaluate user experience using the User Experience Questionnaire Short Version (UEQ) To evaluate satisfaction using the System Usability Scale (SUS) Compare user experience and satisfaction between Shopee and Lazada mobile apps.

Significance of the Study: The user experience of mobile apps and systems is influenced by several critical characteristics, including usability, utility, credibility, findability, attractiveness, accessibility, and value. These factors are critical in establishing how and why users interact with the program and, as a result, contribute to its success (Perlman, 2021). However, not all users have the same demands or behave in the same way. Businesses must establish distinct user profiles and apply UX assessment metrics to properly understand and address consumers' demands (Debnath et al., 2021). The study's findings will assist the following people: Online shopping companies: Companies and stakeholders involved in maintaining mobile B2C platforms would be able to use the assessment and comparison data to improve the user experience of their consumers and the general usability of their apps. This research might help them enhance their platform and improve their company strategy and offerings. UI designers / Mobile developers: Developers would also be aware of the many crucial variables involved in creating a more user-focused business mobile app for a better user experience and more delighted users. They would also be able to identify user experience characteristics that are important to users and can aid in the improvement of their judgments and tactics. Users will have a better experience using the B2C mobile applications that will be reviewed as a result, and their unique demands will be met and given based on their input. Researchers: The study will assist the researchers in expanding their understanding of the notion of user experience and happiness on B2C mobile applications, as well as its significance to specific customers. It would also broaden the researchers' view on the main variables to consider when creating and developing corporate mobile apps based on user input.

Scope and Limitations

This research compares Filipino customers' user experiences and satisfaction with Shopee and Lazada Philippines' mobile applications. The researchers will administer an online survey to 120 people via Google Docs, gathering vital information on the respondents' experiences and levels of satisfaction. The survey only includes consumers who have used the Shopee and Lazada Philippines mobile applications.

Theoretical Framework

The UX focuses on enhancing the relationship between the user and the product by taking hedonic and pragmatic elements into account and understanding and controlling the emotional aspects of product consumption (Ahmad, Hamid, & Lokman, 2021). Multiple studies have explored the link between user experience and customer happiness, concentrating on how these characteristics lead to consumer loyalty, according to Setiyawati and Bangkalang (2022). Their research found that a better user experience boosts consumer happiness. Furthermore, according to Boi-Kudri (2022), good customer satisfaction and perceived value have a direct impact on customer loyalty, with perceived value being a significant predictor of satisfaction. Setiyawati and Bangkalang (2022) underlined the significance of user experience in connection satisfaction as well. Figure 1 shows the relationship between user experience and customer satisfaction.

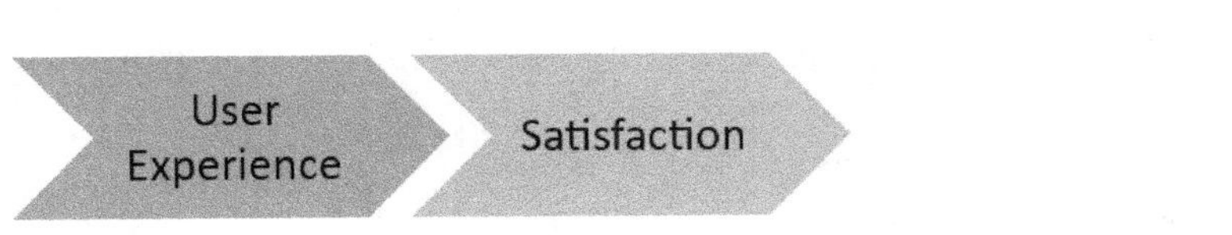

Figure 1. The relationship among user experience and user satisfaction adapted from Setiyawati & Bangkalang (2022, p.2). According to Ninyikiriza et al. (2020), mobile applications for online commerce are only likely to succeed if clients are suitably pleased. Experience satisfaction is one of the most important aspects to consider while building and enhancing a business. Recognizing this impact on e-commerce platforms will lead to the development of mobile B2C apps such as Shopee and Lazada. E-commerce assessment is a critical stage in ensuring the success of these apps. Given the rapid expansion of mobile commerce, it is critical to assess user experiences. The short version of the User Experience Questionnaire (UEQ), System Usability Scale (SUS), and Net Promoter Score (NPS) metrics were used in this study to analyze the user experience and satisfaction of the Shopee and Lazada Philippines mobile applications. This review not only acts as a great resource for developers wanting to improve design and functionality, but it also aids in increasing revenue, improving designs, and meeting customer expectations for these B2C mobile applications.

Figure 2. Method of the study adapted from Setiyawati & Bangkalang (2022, p.2). Respondents Shopee and Lazada Philippines Mobile app user Male = 51; Female = 61 with a total of 120 sample size

Research Methodology

A descriptive-comparative research approach was employed in this study to describe the overall user experience and user satisfaction of Shopee and Lazada Philippines mobile app users. The outcomes of the user experience and user satisfaction evaluations were compared between the two retail systems. User Experience Questionnaire Short Version (UEQ-S) – User Experience The purpose of the UEQ (User Experience Questionnaire) is to give end-users with a quick and simple evaluation tool that provides a full perspective of their experience with a product. The UEQ should allow users to communicate their feelings, thoughts, and attitudes concerning the product under consideration quickly and easily (Schrepp, Hinderks, & Thomaschewski, 2017).

To guarantee its applicability, the UEQ (User Experience Questionnaire) was designed utilizing data analytics (Schrepp, 2015). The UEQ is comprised of six scales totaling 26 items, each assessing a different component of the user experience: Attractiveness evaluates users' impressions of the product. Perspicuity measures the simplicity of becoming acquainted with and learning how to use a product. Efficiency measures a user's capacity to complete tasks with minimal effort. Dependability is a measurement of user influence over the interaction. Stimulation measures the level of excitement and motivation when a product is utilized. Novelty measures the level of innovation and creativity and the product's appeal to users. Perspicuity, Efficiency, and Dependability are pragmatic, goal-oriented quality dimensions, but Attractiveness is a pure valence factor. Stimulation and novelty are hedonic quality features that are not goal-directed (Schrepp, 2015).

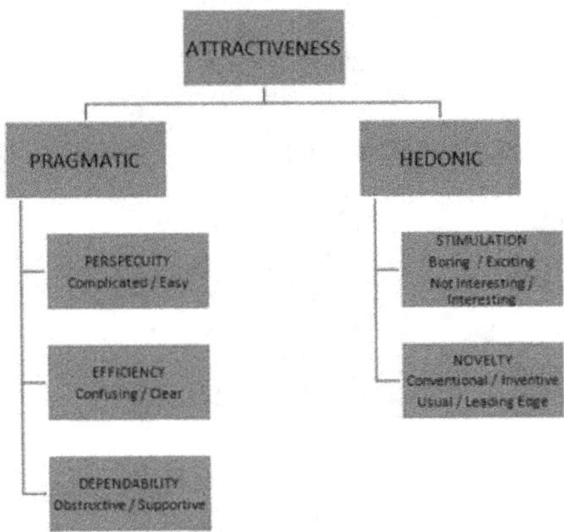

Figure 3.2 User Experience Questionnaire

Survey Administration

Online – Google Docs (survey) that has one survey set with Shopee that comes first and one set survey with Lazada that comes first with 60 unique respondents. Data Analysis used are Descriptive Statistics and T-tests using Microsoft Excel

Results and Discussion

UEQ (User Experience Questionnaire) The Lazada UEQ and Shopee UEQ outcomes are analyzed using the means of the scales' pragmatic and hedonic qualities. These characteristics classify the scales based on the items that were largely rated. Pragmatic Quality is based on three scales: Efficiency, Perspicuity, and Dependability, and Hedonic Quality is based on Stimulation and Novelty. Furthermore, these scales have a total of eight components. According to the accepted interpretation, values on the scale between -0.8 and 0.8 suggest a neural evaluation of the relevant scale, while values more than or equal to 0.8 indicate a positive evaluation, and values less than or equal to -0.8 indicate a negative evaluation. The scales range from -3 (worst) to +3 (outstanding). In real-world applications, however, only numbers within a particular range are often identified. It is extremely rare to discover results more than +2 or less than -2 since the means were calculated among a wide range of individuals with varying opinions and replying patterns.

Table 1.3 Two samples T-Test assuming unequal variances

Qualities	P-Value	Interpretation
Perspicuity	0.727	No Significant Difference
Efficiency	0.032	Significant Difference
Dependability	0.013	Significant Difference
Stimulation	0.006	Significant Difference
Novelty	0.548	No Significant Difference

Table 1.3 displays a comparison of Lazada and Shopee UEQ Results based on the Scales that classified the questionnaire items. The computations were compared to the Alpha or significance level, which is 0.05; if the t-test computation score values are smaller than the Alpha, the null hypothesis (No Significant Difference) is rejected and the alternative hypothesis (Significant Difference) is accepted. Perspicuity has a score of 0.726 based on the two-sample t-test computation

results, suggesting that there is "No Significant Difference" between the two mobile applications on this scale. Efficiency was 0.0318, implying that there is a "Significant Difference" between the two programs. Dependability got a score of 0.0134, implying that there is a "Significant Difference" between the two apps on this scale. Stimulation has a score of 0.0056, suggesting that there is a "Significant Difference" in this scale between the two mobile apps. Finally, the scale Novelty has a score of 0.5482, indicating that there is "No Significant Difference" on both applications when it comes to this scale. Displays a comparison of Lazada and Shopee SUS Results based on the table's SUS ratings. The calculated P value was compared to the Alpha, which represents the likelihood of rejecting the null hypothesis. The null hypothesis (No Significant Difference) is rejected if the P value (probability) is less than the alpha, while the alternative hypothesis (Significant Difference) is accepted if the P value (probability) is less than the alpha. Based on the two-sample t-test computation findings, the P value is 0.000027, which is lower than the rejection area or alpha, showing a "Significant Difference" in the SUS ratings of the two mobile applications.

Conclusion

Finally, in terms of user experience, the findings of the User Experience Questionnaire (UEQ), System Usability Scale (SUS), and Net Promoter Score (NPS) ratings demonstrated a clear preference for Shopee over Lazada. According to the UEQ results, Shopee scored higher marks in Pragmatic and Hedonic quality, reflecting the app's useful and pleasant features. The SUS findings also revealed a substantial difference between the two mobile applications, with Shopee earning a B. Lazada, on the other hand, scored a D. Furthermore, Shopee had a larger number of Promoters, suggesting a better degree of consumer loyalty and happiness, according to the NPS statistics. In terms of user experience, usability, and customer happiness, these findings indicate that Shopee is the superior alternative for users.

Recommendation

In future studies, the researchers advocate increasing the number of respondents to better understand the diverse demands of consumers for B2C mobile applications. Furthermore, this will confirm and deliver more essential data to the Lazada and Shopee platforms. Second, recognizing the unique demands of users based on their B2C mobile application inclinations will help to answer the issue of "what" specific criteria the user expects while using and navigating B2C mobile apps. Finally, it is critical to assess various B2C platforms that require development and feedback on their application. Responses from different users would have a significant influence on the progress of the specific platform itself because meeting the wants of consumers is one of the things that considerably affect and maintain enterprises. Aside from B2C mobile applications, some several more applications and platforms require upgrading. Using the main criteria highlighted in the study will have a significant impact not just on e-commerce platforms, but also on other applications that are heavily impacted by user experience.

References

1. Adobe. (2020, June 5). Usability Metrics: Measuring UX Design Success | Adobe XD. Ideas. Retrieved November 8, 2022, from https://xd.adobe.com/ideas/process/user-testing/usability-metrics-measuring-ux-design-success/

2. Ahmad, N. A. N., Hamid, N. I. M., & Lokman, A. M. (2021). Performing Usability Evaluation on Multi-Platform Based Application for Efficiency, Effectiveness and Satisfaction Enhancement. International Journal of Interactive Mobile Technologies, 15(10).

3. Aquino, M. A. T., Manlapas, L. R. S., Margate, A. M. N., Rivera, P. M. B., & German, J. D. A Usability Study on the Online Retail Platform of Local Pharmacies in the Philippines.

4. Arcega, A. (2017, January 27). Pinoys do most online shopping during office hours, research shows | SciTech |. GMA News Online. Retrieved December 16, 2022, from https://www.gmanetwork.com/news/scitech/technology/597374/pinoys-do-most-online-shopping-during-office-hours-research-shows/story/

5. Baquero, A. (2022). Net Promoter Score (NPS) and Customer Satisfaction: Relationship and Efficient Management. Sustainability, 14(4), 2011.

6. Boichuk, O. (2022, October 4). The Nine Principles of UX Design Psychology: Can You Predict the Behavior of Your Users? UX Magazine. Retrieved December 5, 2022, from https://uxmag.com/articles/the-nine-principles-of-ux-design-psychology-can-you-predict-the-behavior-of-your-users

7. Božić-Kudrić, N. (2022). The impact of personalized user experience on customer satisfaction and loyalty in eCommerce (Doctoral dissertation, University of Zagreb. Faculty of Economics and Business).

8. Bruton, L. (2022, October 13). The 5 elements of UX design explained. UX Design Institute. Retrieved November 6, 2022, from https://www.uxdesigninstitute.com/blog/5-elements-of-ux-design/

9. Centeno, T. A., Pacheco, A. C., Sigua, G. D., Tan, N. J., & Fernandez, R. (2022). Understanding the Influence of Mobile Interface of Shopee and Lazada on Customer Conversion. Journal of Business and Management Studies, 4(2), 192-202.

10. Debnath, N., Peralta, M., Salgado, C., Baigorria, L., Riesco, D., Montejano, G., & Mazzi, M. (2021). Digital transformation: A

quality model based on ISO 25010 and user experience. EPiC Series in Computing, 75, 11-21.

11. Drew, M. R., Falcone, B., & Baccus, W. L. (2018). What does the system usability scale (SUS) measure? validation using think aloud verbalization and behavioral metrics. In Design, User Experience, and Usability: Theory and Practice: 7th International Conference, DUXU 2018, Held as Part of HCI International 2018, Las Vegas, NV, USA, July 15-20, 2018, Proceedings, Part I 7 (pp. 356-366). Springer International Publishing.

12. Escanillan-Galera, K. M. P., & Vilela-Malabanan, C. M. (2019). Evaluating on user experience and user interface (UX/UI) of Enertrapp a mobile web energy monitoring system. Procedia Computer Science, 161, 1225-1232.

13. Huang, J., & Wang, X. (2022). User Experience Evaluation of B2C E-Commerce Websites Based on Fuzzy Information. Wireless Communications and Mobile Computing, 2022.

14. ISO 9241-210:2019(en) Ergonomics of human-system interaction — Part 210: Human-centred design for interactive systems. (2019). ISO. Retrieved November 11, 2022, from https://www.iso.org/obp/ui/#iso:std:iso:9241:-210:ed-2:v1:en

15. Jay Pagkatotohan. (2022, January 17). [Battle of the Brands] Lazada vs Shopee: Which is the Better Online Shopping Site? Retrieved December 16, 2022, from https://www.moneymax.ph/personal-finance/articles/lazada-vs-shopee-review

16. Keenan, M. (n.d.). Global Ecommerce: Stats and Trends to Watch (2022). Shopify Plus. Retrieved November 16, 2022, from https://www.shopify.com/ph/enterprise/global-ecommerce-statistics

17. Keiningham, T. L., Aksoy, L., Cooil, B., Andreassen, T. W., & Williams, L. (2008). A holistic examination of Net Promoter. Journal of Database Marketing & Customer Strategy Management, 15(2), 79-90.

18. Times, T. M. (2022, July 30). Pinoys' online shopping habits continue – study. The Manila Times. Retrieved December 16, 2022, from https://www.manilatimes.net/2022/07/31/business/sunday-business-it/pinoys-online-shopping-habits-continue-study/1852834

19. Yatprom, S. (2021, January 12). How mCommerce is shaping the online commerce in Asia. Scandasia. Retrieved October 8, 2022, from https://scandasia.com/how-mcommerce-is-shaping-the-online-commerce-in-asia/

19. Shoppertainment Unveiled: The Impact of Tiktok Shopping Recommendation Videos on Consumer Behavior of Youth.

Ezekiel U. Manuel, Dba, Allex P. Salmo, James L. Sandalan and Andrea Irish P. Villanueva

Polytechnic University of the Philippines San Juan Branch, San Juan City, **Philippines**

Asbtract: This study explores the impact of TikTok Shopping Recommendation Videos on the Consumer Behavior of Youth (ages 20-24). TikTok, a rapidly growing social media platform, has become a popular discovery tool for consumers due to its diverse shopping recommendation videos, known as "Shoppertainment." Convenience and snowball sampling methods were used, and a survey questionnaire was administered to 100 TikTok users in Barangay Addition Hills, Mandaluyong City. The results show that TikTok significantly influences consumer behavior regarding actions, decisions, product satisfaction, and repurchase intention. Consumers recognize the usefulness of TikTok Shopping Recommendation Videos for discovering new items, making purchase decisions, and appreciating their visual appeal. Respondents trust businesses, content creators, and influencers, impacting their shopping decisions. Prioritizing brand and content creator/influencer suggestions, trending items, and product safety and compliance are essential factors driving repeat purchases. Negative experiences can significantly affect consumers' inclination to repurchase, potentially leading to a loss of trust. This study sheds light on TikTok's profound influence on the behavior of young consumers and provides valuable insights for businesses and marketers seeking to leverage this platform for effective consumer engagement and marketing strategies.

Keywords: TikTok, Shopping Recommendation Videos, Consumer Behavior, Marketing, Shoppertainment

INTRODUCTION

Social media has revolutionized communication and information sharing, becoming an essential aspect of modern life (Lin, 2020). In the Philippines, a prominent player in internet and social media usage, TikTok has gained widespread popularity, evolving into a platform for product promotion and driving unplanned purchases known as "Shoppertainment" (BBC, 2020; Inquirer, 2021). Consumer buying behavior, a critical aspect of consumer decision-making, is significantly influenced by TikTok users who act as influencers (Southern, 2022). Nguyen (2022) emphasizes the impact of influencers on consumer behavior, as trust in their product reviews and recommendations play a pivotal role in decision-making. TikTok's meteoric rise in global popularity, surpassing one billion active users, highlights its significance as a major social media application (Koyak, 2021). Known for its short video content, TikTok serves as a significant platform for accessing and consuming information (Hoi & Yin, 2023).

However, the credibility of influencers on TikTok remains a subject of debate, as not all followers trust their recommendations (Nguyen, 2022). A study in the Philippines by Araujo et al. (2022) underscores emotional, entertaining, and informative factors' influence on consumer behavior, especially concerning TikTok shopping recommendation videos, fostering a positive connection, and influencing purchase intentions. TikTok also impacts consumer decisions through live selling, short-video content, and affiliate marketing (De Jesus & Santiago, 2021), enabling consumers to assess product worthiness based on various personal, social, cultural, and psychological factors. In light of the significant influence of TikTok on consumer behavior, this study aims to explore the impact of TikTok shopping recommendation videos on the buying behavior of youth in Barangay Addition Hills, Mandaluyong City, contributing to the existing knowledge on this topic.

Theoretical Framework

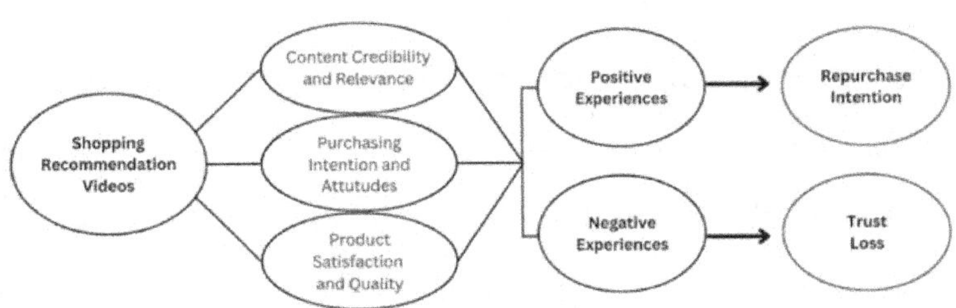

Figure1. Repurchase Intention on Shopping Recommendation Videos

The Repurchase Intention on Shopping Recommendation Videos model, developed by Salmo et al., provides insights into the factors influencing consumers' behavior when encountering shopping recommendation videos. In the scope of this study, the focus is on TikTok shopping recommendation videos, which capture consumers' attention. As consumers watch these videos, they evaluate their interest in the showcased products and make decisions about purchasing them. Positive experiences with the products they buy based on the videos contribute to their continued engagement and repurchase intentions. Negative experiences, on the other hand, can lead to a loss of trust. The model emphasizes the importance of content credibility, relevance, purchasing intention, attitudes, and product satisfaction in shaping consumer behavior in response to shopping recommendation videos.

Conceptual Framework

In this research study, the researchers recognize the importance of establishing a strong foundation based on relevant laws and guidelines. In this regard, Republic Act No. 7394, also known as the Consumer Act of the Philippines, will be utilized as the basis law and guide for the investigation. This Act holds significant relevance in shaping consumer rights and protection, aligning closely with the research objectives. By incorporating this law, the researchers aim to shed light on the intricate relationship between TikTok shopping recommendation videos and consumer behavior while upholding ethical practices and adhering to legal obligations.

Paradigm of the Study:

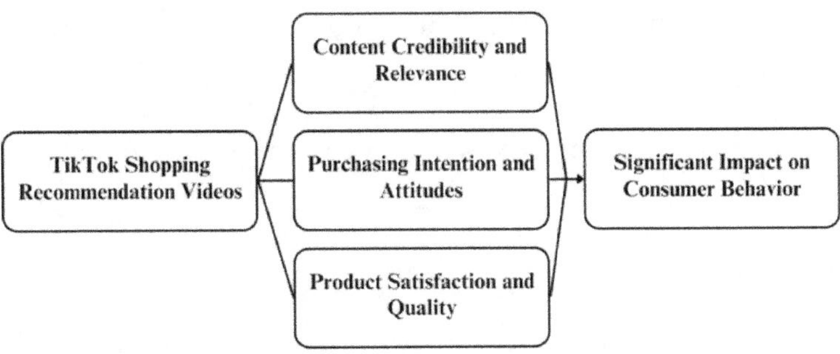

Figure 2. Paradigm of the Study

The research paradigm employed in this study aims to comprehensively investigate the impact of TikTok shopping recommendation videos on the consumer behavior of youth. It delves into aspects such as content credibility and relevance, investigating the perceived trustworthiness and expertise of content creators, influencers, and brands, the accuracy of product information presented, and the alignment of recommendations with viewers' specific needs and preferences. Additionally, the study investigates the impact of these videos on consumers' purchasing intentions and attitudes, analyzing whether they contribute to a heightened interest in the recommended products and influence favorable attitudes towards the featured brands or products. Moreover, the research paradigm extends its focus to the realm of product expectation and satisfaction. It scrutinizes the correlation between TikTok shopping recommendation videos and consumers' satisfaction with the products they purchase. This study seeks to determine the extent to which consumer satisfaction is influenced by video recommendations and whether the perceived quality of the products aligns with their expectations derived from the videos. By considering these factors, the researchers contribute to a deeper understanding of the influence of TikTok videos on consumer behavior and provide valuable insights for both consumers and marketers in the context of the digital age.

Objectives of the Study: This study aims to investigate The Impact of TikTok Shopping Recommendation Videos on the Consumer Behavior of Youth in Barangay Addition Hills, Mandaluyong City. The following are the objectives that the researchers identified in conducting this study:

1. To understand the decisions and actions made by consumers when they purchase a product based on TikTok Shopping Recommendation Videos.

2. To determine whether the products have successfully met consumers' expectations and provided them with satisfaction.

3. To investigate the possibility of repeat purchases among youth influenced by TikTok shopping recommendation videos.

Methodology

Study Design: This Quantitative study used a Descriptive Research Design to investigate the impact of TikTok shopping recommendation videos on the consumer behavior of youth consumers in Barangay Addition Hills, Mandaluyong City. The study focused on describing and understanding the factors influencing purchasing decisions, the expectation and satisfaction with the products purchased, and investigating the possibility of repeat purchases. The findings of this descriptive research will contribute valuable insights into how TikTok shopping recommendation videos affect the consumer behavior of youth in Barangay Addition Hills, Mandaluyong City. **Sample/Population of the Study** This study focuses on youth residing in Barangay Addition Hills, Mandaluyong City, as the target population. The specific age group is between 20 to 24 years old, as they are the age group that purchases products based on watching TikTok shopping recommendation videos. Additionally, the respondents are three (3) years active TikTok users and consumers and have been influenced multiple times by watching TikTok shopping recommendation videos. This study has 100 respondents, and the sampling methods used to recruit the participants were Convenience and Snowball Sampling. Convenience sampling was used to recruit the respondents because the target population is near the researchers. Snowball sampling, on the other hand, was used for existing respondents to refer additional participants who fit the criteria of this study. **Data Gathering Tools:** The data collected from the respondents were gathered using a combination of researcher-developed and adapted-made questionnaires. Some questions were adapted from a study related to the topic that was conducted by Araujo et al. (2022). The survey was distributed via Google Forms and divided into five (5) categories related to the study's objectives. It contains 25 questions and four (4) options for answering the questionnaires: Agree, Strongly Agree, Disagree, and Strongly Disagree.

Treatment of Data

4-POINT LIKERT SCALE

Verbal Interpretation	Numerical Allocation
Strongly Disagree	1.00 – 1.75
Disagree	1.76 – 2.50
Agree	2.51 – 3.25
Strongly Agree	3.26 – 4.00

Figure 4. 4-point Likert scale

The 4-point Likert scale used in the survey questionnaires collected from the selected respondents is shown in the table above. This scale reflects the choices made by respondents while answering questions, as well as the numerical allocation that served as the basis for the average response. The data gathered was encoded, tallied, computed, and analyzed. The statistical tools that researchers used are the weighted mean and verbal interpretation.

Ethical Considerations

The researchers submitted a Letter of Intent to the Vice President for Research, Extension, Planning, and Development at the Polytechnic University of the Philippines Main Campus for Ethics Clearance. This letter included a copy of the survey questionnaire, consent forms, etc., for them to review to guarantee that the study and survey questions were not harmful to any respondents. This process sought to build trust between the researchers and the respondents to ensure their inclusion in the study. Before participating, the respondents were given an informed consent letter saying they willingly granted consent. The researchers complied with Republic Act 10173, also known as the Data Privacy Act of 2012, which ensured that all information provided by respondents was kept strictly confidential and utilized only for research purposes.

Results and Discussions

This chapter presents the results and the discussion of the study. It consists of presenting, analyzing, and interpreting the results that address the study's objectives.

Table 1: *Respondents' Assessment of the Reasons Why Consumers View TikTok Shopping Recommendation Videos*

Indicators	Rank	Weighted Mean	Verbal Interpretation
1. I watch TikTok Shopping Recommendation Videos because they are visually appealing.	2	3.30	Strongly Agree
2. I find TikTok Shopping Recommendation Videos to be a	1	3.50	Strongly Agree

great way to discover new products or brands.			
3. TikTok Shopping Recommendation Videos have led me to choose the best products.	4	3.24	Agree
4. I find TikTok Shopping Recommendation Videos helpful in making purchasing decisions.	3	3.28	Strongly Agree
5. I watch TikTok Shopping Recommendation Videos whenever there is an exclusive limited packaging, product/ or promo.	5	3.17	Agree

<table>
<tr><td>OVERALL WEIGHTED MEAN</td><td>3.30</td><td>Strongly Agree</td></tr>
</table>

Based on the result above, the consumers recognize TikTok Shopping Recommendation Videos and strongly agree with their usefulness in discovering new items, making purchase decisions, and appreciating the visual attractiveness of every shopping recommendation video they watch. In relation to Rank 1, which is Indicator 2, Jarboe (2022) stated that TikTok has a significant and positive influence throughout the entire purchasing process. It functions as a word-of-mouth marketplace where post-purchase actions drive engagement, with users relying on brands, creators, and trending topics to explore new products. The research reveals that 50% of TikTok users experience feelings of joy, excitement, or happiness regarding their purchases. Furthermore, 58% of users discover new brands and products directly on the platform, and 44% are prompted to make immediate purchases. Notably, TikTok is a discovery platform 1.1 times more frequently than other social media platforms. The table above implies that consumers in TikTok Shopping Recommendation Videos typically believe that businesses, content creators, and influencers impact their shopping decisions. They trust their recommendations, believe in the product quality, and believe that having suggested things brings them closer to them. However, content creators/influencers have a somewhat smaller influence than companies, and there is a modest amount of confidence in their suggestions when compared to traditional advertising. In relation to Rank 1, which is Indicator 1, as discussed by Southern (2022), it was found that a significant 44% of TikTok users actively discover new products through videos posted by various brands on the platform. This statistic highlights the influential role of TikTok as a key medium for users when it comes to discovering and subsequently purchasing new products. The data suggest that TikTok serves as an effective channel for brand exposure and product exploration, as users engage with brand-related content and are enticed to explore and make purchases based on the videos they come across.

Table 3

Respondents' Assessment of the Decisions Made By Consumers before Buying a Product

Indicators	Rank	Weighted Mean	Verbal Interpretation
1. I do research first before buying a product that I watched on TikTok.	3	3.53	Strongly Agree
2. I check the price first before buying a product that I watched on TikTok.	2	3.59	Strongly Agree
3. I double-check the video first so that I can see if the	1	3.61	Strongly Agree

| 4. I don't make decisions – If I like something I watch on TikTok, I will buy it. | 4 | 2.47 | Disagree |

| **OVERALL WEIGHTED MEAN** | | 3.30 | **Strongly Agree** |

The results above suggest that consumers are highly proactive in their decision-making process when purchasing products they watched on TikTok. They prioritize conducting research, checking prices, and reviewing videos to ensure that the product meets their expectations before making a purchase. They value making informed decisions rather than impulsive purchases. In relation to Rank 1, which corresponds to Indicator 3, the research conducted by De Jesus and Santiago (2022) highlights TikTok's significant impact on the customer experience and consumer behavior. The study indicates that TikTok has emerged as a platform experiencing robust growth, making it imperative for businesses to actively engage with this dynamic platform. Furthermore, the research reveals that consumers on TikTok display a discerning approach toward their purchasing decisions. They often double-check the credibility and reliability of the videos they encounter on TikTok before making a purchase. This cautious behavior demonstrates their desire to ensure that their hard-earned money is not wasted on subpar or unsatisfactory products. The result above demonstrates that consumers usually agree with significant elements of their TikTok product experiences. While there are cases where things set off from expectations or arrive damaged despite favorable comments, buyers generally feel that the products obtained are consistent with what they saw on TikTok and that the cost is appropriate. In relation to Rank 1, specifically Indicator 1, Lova and Budaya (2023) have asserted that customer satisfaction is closely intertwined with customer attitudes and intentions. These aspects are integral parts of customer behavior and directly impact customers' positive behavioral intentions. Indicator 2, which is also part of Rank 1, Rochman and Kusumawati's (2023) study, aims to analyze the effects of promotions, influencers, convenience, service quality, and price. Price, in particular, emerges as a crucial determinant in purchasing a product. Consumers tend to compare prices of identical products and are inclined to opt for a lower-priced option, considering economic factors and the perceived quality of the product relative to its price. Based on the result above, it indicates that negative experiences can significantly impact their willingness to repurchase a product. The findings reveal that customers largely agree on the numerous criteria driving their repeat purchases. They prioritize brand and content creator/influencer suggestions, trending items, and product safety and compliance. Negative experiences, on the other hand, might have a major influence on their inclination to buy a product. In relation to Rank 1, specifically Indicator 4, the research conducted by Wulf and Köcher (2017) focused on examining the effects of product certifications, specifically a partial certification strategy, on consumers' perceptions. This research sheds light on the importance of certifications in providing assurance and influencing consumer perceptions. Building upon this, it is crucial for products to be certified, verified, safe, and compliant with health standards for various reasons. The findings of Wulf and Köcher's study align with the broader significance of certifications. Certifications play a vital role in establishing trust and confidence among buyers by ensuring that products meet specific standards and have undergone rigorous testing or evaluation processes.

Conclusions

In conclusion, this study has provided empirical evidence supporting the significant impact of watching shopping recommendation videos on TikTok on consumer behavior. By uncovering and analyzing consumers' behaviors and decisions during the purchase process, the study has shed light on the underlying factors that drive their engagement with TikTok shopping recommendation videos. Specifically, it has identified the reasons behind consumers' interest in these videos, the impact of brands and content creators/influencers, and the decision-making process preceding a product purchase. These findings underscore the importance of TikTok as a valuable platform within consumers' decision-making journeys. Moreover, the study has revealed that consumers who had positive experiences and found their expectations met through products purchased based on TikTok shopping recommendations reported high levels of satisfaction.

Conversely, instances of negative experiences deterred consumers from relying on TikTok shopping recommendations for future purchases. These findings contribute to the existing body of knowledge on the influence of social media platforms, such as TikTok, in shaping consumer behavior and highlight the need for businesses to strategically utilize this platform to effectively engage with and influence their target audience.

References

Araujo et al. (2022). *Influence of TikTok Video Advertisements on Generation Z's Behavior and Purchase Intention.* Retrieved from International Journal of Social and Management Studies Website: https://ijosmas.org/index.php/ijosmas/article/view/123

Baclig, C.E. (2022). *Social media, internet craze keep PH on top 2 of world list.* Retrieved from the Inquirer website: https://newsinfo.inquirer.net/1589845/social-media-internet-craze-keep-ph-on-top-2-of-world-list

Barcelona et al. (2022). *#Budolfinds: The Role of TikTok's Shopee Finds' Videos in the Impulsive Buying Behavior of Generation Z Consumers.* Retrieved from the Research Gatewebsite: https://www.researchgate.net/publication/365353518_Budolfinds_The_Role_of_TikTok's_Shopee_Finds'_Videos_in_the_Impulsive_Buying_Behavior_of_Generation_Z_Consumers

BBC. (2020). *How TikTok changed the world in 2020*. Retrieved from the BBC website: https://www.bbc.com/culture/article/20201216-how-tiktok-changed-the-world-in-2020

Boston University School of Public Health. (2022). *The Social Cognitive Theory*. Retrieved from the Boston University School of Public Health website: https://sphweb.bumc.bu.edu/otlt/mph-modules/sb/behavioralchangetheories/behavioralchangetheories5.html

Ceci, L. (2023). *TikTok – Statistics & Facts*. Retrieved from the Statista website:

https://www.statista.com/topics/6077/tiktok/#topicOverview

De Jesus, F. and Santiago, S. (2021). *TikTok as an Advertising Platform: Its Effectiveness To Young Adult Consumers Buying Decision Making in the 4th District Of Nueva Ecija*. Retrieved from The Law Brigade Publishers: https://thelawbrigade.com/wp-content/uploads/2022/12/Fhrizz-Shirley-AJMRR.pdf

Estella, P. and Löffelholz, M. (n.d.). *Social Networks*. Retrieved from the Website of Media Landscapes: https://medialandscapes.org/country/philippines/media/social-networks

Hoi, N.K. and Yin, L.K. (2023). *Short Videos, Big Decisions: A Preliminary Study of Tik Tok's Role in E-Commerce Consumer Behaviour*. Retrieved from the European Journal of Business and Management Research website: https://www.ejbmr.org/index.php/ejbmr/art icle/view/1951

INQUIRER.net. (2021). *TikTok predicts Shoppertainment to dictate purchase trends in PH mega sales*. Retrieved from the Inquirer website: https://technology.inquirer.net/110935/tiktok-predicts-shoppertainment-to-dictate-purcha se-trends-in-ph-mega-sales

Jarboe, G. (2022). *Research Reveals TikTok's Impact on Consumers' Purchase Journeys*. Retrieved from the Search Engine Journal website: https://www.searchenginejournal.com/tiktoks-impact-purchasing-research/453960/#close

Koyak, B. (2021). *What is TikTok?*. Retrieved from the Laurus College website: https://lauruscollege.edu/meet-tiktok/

Lin, J. (2020). *Social Media has changed the lives of Modern Society*. Retrieved from the Summit News website: https://summitpsnews.org/2020/03/24/social-media-has-changed-the-lives-of-modern-society/

Lova, A.N. and Budaya, I. (2023). *Behavioral of Customer Loyalty on E-Commerce: The Mediating Effect of E-Satisfaction in Tiktok Shop*. Retrieved from the Journal fo Scientific, Research, Education, and Technology website: https://jsret.knpub.com/index.php/jrest/article/view/43/33

Nguyen, T. (2022). *The Impact of TikTok Influencer Marketing on Consumer Behavior*. Retrieved from the Theseus website: https://www.theseus.fi/handle/10024/748519

Ohio University. (2022). *5 Consumer Behavior Theories Every Marketer Should Know*. Retrieved from the Ohio University website:

https://onlinemasters.ohio.edu/blog/con sumer-behavior-theories-every-marketer-should-know/

Republic Act No. 7394: The Consumer Act of the Philippines. (1992). Retrieved from the Official Gazette of the Philippines Website: https://www.officialgazette.gov.ph/199 2/04/13/republic-act-no-7394-s-1992/

Rochman, H.N. and Kusumawati, E. (2023). *Analysis of the influence of promotions, influencers, convenience, service quality and prices on the Tiktok application on purchasing decisions on the "Tiktok Shop."* Retrieved from the International Journal of Latest Research in Humanities and Social Sciences: http://www.ijlrhss.com/paper/volume-6-issue-4/2-HSS-1843.pdf

Southern, M.G. (2022). *TikTok A Key Part of Consumers' Path To Purchase.* Retrieved from the Search Engine Journal website: https://www.searchenginejournal.com/tiktok-a-key-part-of-consumers-path-to-purchase/437912/#close

University of Delaware (n.d.). *Chapter 6. Consumer Buying Behavior Notes.* Retrieved from the website of University of Delaware: https://www1.udel.edu/alex/chapt6.html

Wulf, L. and Köcher, S. (2017). *The Bright and Dark Sides of Product Certification: Exploring Side Effects on Consumers' Perceptions of Non-Certified Products: An Abstract.* Retrieved from Springer Nature website: https://link.springer.com/chapter/10.1007/978-3-319-66023-3_174

20. System for Love Healthcare for Loved Ones based on IoT.

Dr. Kazi Kutubuddin Sayyad Liyakat

Professor, Department of Electrical Engineering,

Brahmdevdada Mane Institute of Technology, Solapur (MS), **India**

Abstract- Love Heath remain built on feelings alike Love and liking. The feelings may originate based on in what way firm or gently individuals become to recognize individuals. It is possible for someone's Love or devotion to not always be returned. One of the most wonderful human feelings emerges when love is shared. Whenever love doesn't get returned, people feel rejected and miserable. It has implemented a lot of patience and diligence to research Love, liking, and unrequited Love. When does adoration turn into Love, for example, is one of the key questions that certain researches have focused on responding. Is there a fundamental difference between the emotions of love and liking? And what happens when someone experiences unrequited love? Love can occur between loved couples, brothers, kids and their parents, parents, and their children's friends, spouses, and parents and wives. A mechanism is needed to demonstrate their affection for one another and to honor loved ones. So, to demonstrate the love memory, we will employ current methods. No such system was ever conceived or present in the past. We develop a system which is used to demonstrate the love health for a persons who live a part. Love heathcare is important aspect for a loved ones as like healthcare or heartcare.

Keywords: - Love; Remembrance; IoT; Lamp; Internet; Wi-Fi; Healthcare;

Introduction

LOVE is a strong, sincere bond with another individual like lovers, wife, father-mother, brother-sister, child etc. According to Thiago de Almeida (2020)[1-3], being deeply attached to someone is another way to describe love[4]. Love might be defined as a strong or intense fondness for somewhat[5-7]. Love has numerous different connotations, both as verb or a noun. Numerous sources identify more forms of love, but four seems like a good starting point.[8-10] Some metaphors for love health which could denote varying levels of intensity or closeness include fondness, dedication, affection, and adoration. How do you depict love health and love traditionally?[11] Feelings like love, liking, etc. are the foundation of intimate relationships. As individuals get to know one another, their emotions could rise to the surface quickly or develop more gradually[12]. Other times, Rahmat Kochhar (2015) does not reciprocate sentiments of love or affection. When love is shared, one of the most romantic feelings a person can have occurs. When someone doesn't love them back, they feel forbidden and dejected. Researchers have worked hard to learn about themes like love, liking, and unrequited love[13-16]. This exploration has engrossed on answering certain fundamental questions like the ones listed below: When may alike turn into love? Is the feeling of love clearly distinct from the feeling of liking? How do individuals feel since they don't feel loved in return? To maintain a positive connection, we need to find the right balance between closeness and separation. Lack of space might make you feel constrained and bored with life. On the other side, being too far away from your mate might make you feel alone and detached. Over time, your sentiments may change, and you two can move apart. The techniques developed by [17-22] for communicating love in long-distance partnerships. However, such a relationship is never feasible. Therefore, a gadget is needed to display the memory of a loved one. Therefore, we are creating a heart-shaped love gadget. The gadget, a heart-shaped lamp called the "love lamp," connects through the internet. The LOVE connection gadget was developed by [56,] and is based on Internet of Things (IoT)[23-30].

"IoT"[13] is an abbreviation for "Internet of Things"[31]. The term "IoT" is derived from the words "Internet"[32] and "Things"[33] with "Things" referring to any internet-connected device [34]. IoT is a system of interconnected algorithms[35-40], machinery, items, quantities, or people, respectively having a unique identification[41-47] and the capability to interconnect information thru a network deprived of the need for human-to-machine interaction. Sensors[248-50] and actuators[51] have been added to IoT in recent years. The primary goal of this study is to create an entirely computerized IoT-based sub-station health monitoring system that allows connected equipment to be secured and monitored from anywhere in the globe by authorised employees at a very low cost[31-35]. While developing a remembrance system, reliability and privacy using IoT technologies are also top priorities [36-40].

Proposed System

Figure 2 beneath shows our suggested setup, while Figure 1 depicts our suggested lamp form. The illustration is only a rough idea that will eventually be put into practice. Net and Wi-Fi are used to link these two bulbs together. For interaction, every lamp needs a unique IP address[41] that is saved in its internal memory[42].

Figure 1- Proposed Lamp Shape

Figure 2- Sample look of proposed device

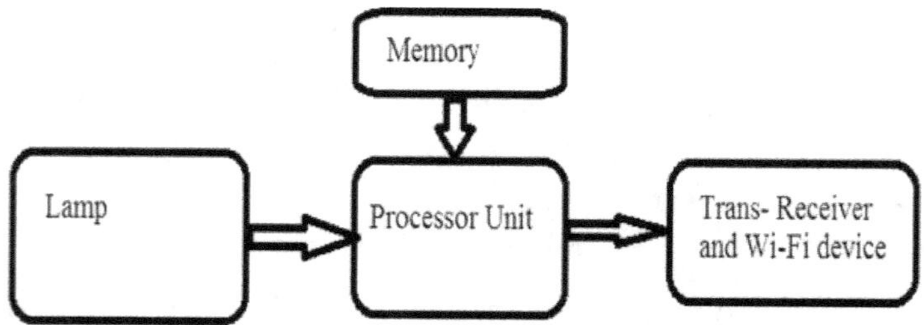

Figure 3- Block Configuration of system

The IP address for the second piece of equipment is kept in the storage unit. Each side has the exact same arrangement installed.

Discussion

Love health is important for the people who live apart from each other and will difficult to meet for long time, so this device made connections and keep love health healthy. These two lights are connected via internet access and Wi-Fi. Each bulb has a distinct IP address which is kept in its internal storage for communication purposes. The IP can be obtained when needed, or as a way to remember someone you love. The suggested system operates as shown in diagram 4 below. Whenever the light button is turned on, the processing unit processes the necessary data and sends it to a different device over Wi-Fi and via the Internet. An other gadget will detect the data obtained over WiFi and the net, and that Lamp will illuminate. This demonstrates the Loved One's memory.

Conclusion

Here, we employ a contemporary method utilizing IoT, i.e., the utilization of Wi-Fi and the Internet. With the help of this technology, you can remember someone without really phoning them. This is beneficial for all family members. Figure 4 depicts the suggested system in operation. In the past, there has never been a system or mechanism for love and remembering. The heart shape was chosen because it represents love, care and affection. Love healthcare is important aspect of loved ones. Love health is important for the people who live apart from each other and will difficult to meet for long time, so this device made connections and keep love health healthy. Since this gadget is only created in pairs, the connection between the two bulbs is kept. We stipulated that no one would add a third bulb to the system. It will alter in the future by utilising sensors.

Reference

[1]. Thiago de Almeida José Fernando Bittencourt Lomônaco, "The concept of love: an exploratory study with a sample of young Brazilians", International Journal of Advanced Engineering Research and Science, 2020, Vol 7, issue 3, pp. 239-257

[2]. Rahmat Kochhar Daisy Sharma, "Role of Love in Relationship Satisfaction", The International Journal of Indian Psychology, 2015, Vol 3, issue 1, pp. 81-107

[3]. Thiago de Almeida José Fernando Bittencourt Lomônaco, "The concept of love: an exploratory study with a sample of young Brazilians", International Journal of Advanced Engineering Research and Science, 2020, Vol 7, issue 3, pp. 239-257

[4]. Ihua M. Ackerman ,Vladas Griskevicius, "Let's Get Serious: Communicating Commitment in Romantic Relationships" ,

Journal of Personality and Social Psychology, 2011, Vol 100, issue 6, pp. 1079-1094

[5].   Ravi Aavula, Amar Deshmukh, V A Mane, et al, "Design and Implementation of sensor and IoT based Remembrance system for closed one", Telematique, 2022, Vol 21, Issue 1, pp. 2769 - 2778.

[6].   Kazi Kutubuddin S. L., "A novel Design of IoT based 'Love Representation and Remembrance' System to Loved One's", Gradiva Review Journal, 2022, Vol 8, Issue 12, pp. 377 - 383.

[7].   Divya Swami, et al, "Sending notification to someone missing you through smart watch", International journal of information Technology & computer engineering (IJITC), 2022, Vol 2, issue 8, pp. 19 - 24

[8].   Halli U M, "Nanotechnology in IoT Security", Journal of Nanoscience, Nanoengineering & Applications, 2022, Vol 12, issue 3, pp. 11 – 16

[9].   Wale Anjali D., Rokade Dipali, et al, "Smart Agriculture System using IoT", International Journal of Innovative Research In Technology, 2019, Vol 5, Issue 10, pp.493 - 497.

[10]. Kazi K. S., "Significance And Usage Of Face Recognition System", Scholarly Journal For Humanity Science and English Language, 2017, Vol 4, Issue 20, pp. 4764 - 4772.

[11]. Miss. A. J. Dixit, et al, "Iris Recognition by Daugman's Method", International Journal of Latest Technology in

Engineering, Management & Applied Science, 2015, Vol 4, Issue 6, pp 90 - 93.

[12]. Miss. Priyanka M Tadlagi, et al, "Depression Detection", Journal of Mental Health Issues and Behavior (JHMIB), 2022, Vol 2, Issue 6, pp. 1 - 7

[13]. Mrunal M Kapse, et al, "Smart Grid Technology", International Journal of Information Technology and Computer Engineering, Vol 2, Issue 6

[14]. Satpute Pratiskha Vaijnath, Mali Prajakta et al. "Smart safty Device for Women", International Journal of Aquatic Science, 2022, Vol 13, Issue 1,

[15]. Waghmare Maithili, et al, "Smart watch system", International journal of information Technology and computer engineering (IJITC), 2022, Vol 2, issue 6, pp. 1 - 9.

[16]. Kazi K S L, "Significance of Projection and Rotation of Image in Color Matching for High-Quality Panoramic Images used for Aquatic study", International Journal of Aquatic Science, 2018, Vol 09, Issue 02, pp. 130.

[17]. Halli U.M., "Nanotechnology in E-Vehicle Batteries", International Journal of Nanomaterials and Nanostructures. 2022; Vol 8, Issue 2, pp. 22.

[18]. Kazi K S, " Detection of Malicious Nodes in IoT Networks based on Throughput and ML", Journal of Electrical and Power System Engineering, 2023, Volume-9, Issue 1, pp. 22- 29.

[19]. Karale Nikita, Jadhav Supriya, et al, "Design of Vehicle system using CAN Protocol", International Journal of Research in Applied science and Engineering Technology, 2020, Vol 8, issue V, pp. 1978 - 1983

[20]. K. Kazi, "Lassar Methodology for Network Intrusion Detection", Scholarly Research Journal for Humanity science and English Language, 2017, Vol 4, Issue 24, pp.6853 - 6861.

[21]. Salunke Nikita, et al, "Announcement system in Bus", Journal of Image Processing and Intelligent remote sensing, 2022, Vol 2, issue 6

[22]. Madhupriya Sagar Kamuni, et al, "Fruit Quality Detection using Thermometer", Journal of Image Processing and Intelligent Remote Sensing, 2022, Vol 2, Issue 5.

[23]. Shweta Kumtole, et al, " Automatic wall painting robot Automatic wall painting robot", Journal of Image Processing and Intelligent remote sensing, 2022, Vol 2, issue 6

[24]. Kadam Akansha, et al, "Email Security", Journal of Image Processing and Intelligent remote sensing, 2022, Vol 2, issue 6

[25]. Shreya Kalmkar, Afrin, et al., " 3D E-Commers using AR", International Journal of Information Technology & Computer Engineering (IJITC), 2022, Vol 2, issue 6, pp. 18-27

[26]. K. K., "Multiple object Detection and Classification using sparsity regularized Pruning on Low quality Image/ video with Kalman Filter Methodology (Literature review)", 2022

[27]. M Pradeepa, et al, "Student Health Detection using a Machine Learning Approach and IoT", 2022 IEEE 2nd Mysore sub section International Conference (MysuruCon), 2022.

[28]. Waghmode D S , et al, "Voltage Sag mitigation in DVR based on Ultra capacitor", Lambart Publications. 2022, ISBN – 978-93-91265-41-0

[29]. Prof. Vinay S , et al, "Multiple object detection and classification based on Pruning using YOLO", Lambart Publications, 2022, ISBN – 978-93-91265-44-1

[30]. Kazi K S, "IoT based Healthcare system for Home Quarantine People", Journal of Instrumentation and Innovation sciences, 2023, Vol 8, Issue 1,

[31]. Rajesh Maharudra Patil " Modelo De Apariencia Discriminatorio Para Un Sólido Seguimiento En Línea De Múltiples Objetivos", Telematique, 2023, Vol 22, Issue 1, pp. 24- 43

[32]. Miss. Mamdyal, Miss. Sandupatia, et al, " GPS Tracking System", International Journal of Advanced Research in Science, Communication and Technology (IJARSCT), 2022, 2, pp. 2492 - 2529

[33]. Kazi K S L, "IoT-based weather Prototype using WeMos", Journal of Control and Instrumentation Engineering, 2023, Vol 9, Issue 1, pp. 10 - 22

[34]. Ravi A. , et al, "Pattern Recognition- An Approach towards Machine Learning", Lambert Publications, 2022, ISBN- 978-93-91265-58-8

[35]. Kazi K S, "IoT-Based Healthcare Monitoring for COVID-19 Home Quarantined Patients", Recent Trends in Sensor Research & Technology, 2022, Vol 9, Issue 3. pp. 26 – 32

[36]. Gouse Mohiuddin Kosgiker, "Machine Learning- Based System, Food Quality Inspection and Grading in Food industry", International Journal of Food and Nutritional Sciences, 2018, Vol 11, Issue 10, pp. 723- 730

[37]. U M Halli, Voltage Sag Mitigation Using DVR and Ultra Capacitor. Journal of Semiconductor Devices and Circuits. 2022; 9(3): 21–31p.

[38]. K. K. S. Liyakat, "Detecting Malicious Nodes in IoT Networks Using Machine Learning and Artificial Neural Networks," *2023 International Conference on Emerging Smart Computing and Informatics (ESCI)*, Pune, India, 2023, pp. 1-5, doi: 10.1109/ESCI56872.2023.10099544.

[39]. Narender Chinthamu, M. Prasad, "Self-Secure firmware model for Blockchain-Enabled IOT environment to Embedded system", Eur. Chem. Bull., 2023, 12(S3), pp. 653 – 660. DOI:10.31838/ecb/2023.12.s3.075

[40]. Vahida Kazi, et al, " Deep Learning, YOLO and RFID based smart Billing Handcart", Journal of Communication Engineering & Systems, 2023, 13(1), pp. 1-8

[41]. Altaf Osman Mulani, Rajesh Maharudra Patil "Discriminative Appearance Model For Robust Online Multiple Target Tracking", Telematique, 2023, Vol 22, Issue 1, pp. 24- 43

[42]. Karale Aishwarya A, et al, "Smart Billing Cart Using RFID, YOLO and Deep Learning for Mall Administration", International Journal of Instrumentation and Innovation Sciences, 2023, Vol 8, Issue- 2.

[43]. K. Kazi, "Systematic Survey on Alzheimer (AD) Diseases Detection", 2022

[44]. K. Kazi, "A Review paper Alzheimer", 2022

[45]. K. K., "Multiple object Detection and Classification using sparsity regularized Pruning on Low quality Image/ video with Kalman Filter Methodology (Literature review)", 2022

[46]. Sultanabanu Kazi, et al.(2023). Fruit Grading, Disease Detection, and an Image Processing Strategy, Journal of Image Processing and Artificial Intelligence, 9(2), 17-34.

[47]. Sultanabanu Kazi, Mardanali Shaikh, Kazi Kutubuddin "Machine Learning in the Production Process Control of Metal Melting" Journal of Advancement in Machines, Volume 8 Issue 2 (2023)

[48]. Liyakat, K.K.S. (2023). Machine Learning Approach Using Artificial Neural Networks to Detect Malicious Nodes in IoT Networks. In: Shukla, P.K., Mittal, H., Engelbrecht, A. (eds) Computer Vision and Robotics. CVR 2023. Algorithms for

Intelligent Systems. Springer, Singapore. https://doi.org/10.1007/978-981-99-4577-1_3

[49]. Kazi Kutubuddin Sayyad Liyakat, "IoT based Smart HealthCare Monitoring", In: Rhituraj Saikia (eds), Liberation of Creativity: Navigating New Frontiers in Multidisciplinary Research, Vol. 2, July 2023, pp. 456- 477, ISBN: 979-8852143600

[50]. Kazi Kutubuddin Sayyad Liyakat, "IoT based Substation Health Monitoring", In: Rhituraj Saikia (eds), Magnification of Research: Advanced Research in Social Sciences and Humanities, Volume 2, October 2023, pp. 160 – 171, ISBN: 979-8864297803

[51]. Priya Mangesh Nerkar1 , Sunita Sunil Shinde, et al, "Monitoring Fresh Fruit and Food Using Iot and Machine Learning to Improve Food Safety and Quality", Tuijin Jishu/Journal of Propulsion Technology, Vol. 44, No. 3, (2023) , pp. 2927 – 2931

https://propulsiontechjournal.com/index.php/journal/article/view/914

21. Transhumanism: A study in Mary Shelley's *Frankenstein*

Pankaj Goswami

Independent Scholar, Former Post Graduate Student, Skb University, Purulia, West Bengal, **India**

Abstract: At the very core of Mary Shelley's Frankenstein, the Creature meets his maker with the narrative of his miserable life, and entreats him to make a female companion with whom he can share his life. Although Victor admits to having been moved by the Creature's eloquence and fine sensations, he reluctantly succumbs to his plea only to destroy the female before completing her, afraid that this new species might pose a threat to the survival of his own. In the encounter of these two species, however, only one seems to have truly "met" the other: the Creature has indeed become with his maker in a way that Victor fails. Given that the dominant narrative point of view up until that moment had been Victor's, readers of the novel have the opportunity of having their ignorance enriched regarding the Creature straight from the Other's mouth, this multiple narrative thus enabling them to take Victor's creation as far more than the monster he sees. Indeed, I would argue that readers do "meet" the Creature while his creator cannot. Taking this central part of the novel as a starting point, this essay will explore the transhuman discourses in Mary Shelley's *Frankenstein*. In expressing his desire to create an improved species,Victor echoes the transhuman discourses of improvement of the human race.

Keywords- Transhumanism, Posthuman, Science fiction, Gothic, Mary Shelley

Introduction

From the very ancient times human beings have been looking for a sorcerer's stone to overcome death and illness. Gradually the development of scientific research and technology, the methods to prolong life and cheat death altered, but the idea remains the same. Still human being can't avoid the death. The craving for eternity is still as strong as million years ago, and in search of an ultimate decision people spend days and nights , years after years of their lives,but the question stays open. Today, it is clear that maintaining life at the cost of moral and ethical values is often imprudent. However, the research goes on giving birth to various biotechnologies that are intended to reveal the key secret of life and master the ability to control life on earth. It goes without saying that debates do not calm down, and the opponents propose a number of counter arguments to transhumanist ideas. Apart from scholar research, there is a wide variety of art in which transhumanist subject is challenged.The novel *Frankenstein* tells about a terrifying experiment performed by a whimsical scientist Victor Frankenstein. Frankenstein's experiment resulted in the creation of a monster that is yet able to love and suffer as well as a naturally born human being. According to the critics, the story told by Shelley is an eloquent example of how biotechnologies provoke the creation of transhuman, often objectified and socially segregated. Hence, current study is intended to retrace how the cautionary twist for transhumanism is introduced in Mary Shelley's novel Frankenstein (1818). Transhumanism is "an international cultural and intellectual movement with an eventual goal of fundamentally transforming the human condition by developing and making widely available technologies to greatly enhance human intellectual, physical, and psychological capacities" (Burdett 13). A common feature of transhumanism and philosophical posthumanism is the future vision of a new intelligent species, into which humanity will evolve and eventually will supplement or supersede it. Transhumanism stresses the evolutionary perspective, including sometimes the creation of a highly intelligent animal species by way of cognitive enhancement. "Transhumanists believe in the perfectibility of the human, seeing the limitations of the human body (biology) as something that might be transcended through technology so that faster, more intelligent, less disease-prone, long-living human bodies might one day exist on Earth"(P.K Nayar, 16). Transhumanism can be viewed as an extension of humanism, from which it is partially derived. Humanists believe that humans matter, that individuals matter. We might not be perfect, but we can make things better by promoting rational thinking, freedom, tolerance, democracy, and concern for our fellow human beings. Transhumanists agree with this but also emphasize what we have the potential to become. Just as we use rational means to improve the human condition and the external world, we can also use such means to improve ourselves, the human organism. In doing so, we are not limited to traditional humanistic methods, such as education and cultural development. We can also use technological means that will eventually enable us to move beyond what some would think of as "human". Is there a more natural state than birth? Our planet's formation. A human being's birth. The cycle of birth, life, and death is the cornerstone of human existence. You may even argue that God gave us a soul because the gift of immortality would be considered excessive. All of that is changing now, thanks to advances in science and technology.Throughout history, there has been a curiosity in improving human beings. To overcome the basic fault in the human lifecycle—aging, the brutal decline of one's faculties, and, eventually, death. Perhaps humanity's greatest unfinished dream is to replicate ourselves as something brighter, stronger, and immune to the ravages of time. To advance from common mortals to God-status. From a scientific and technological standpoint, this form of pseudo-immortality is referred to as "transhumanism": the biotechnological improvement of people that essentially eliminates the biological terminal frailties of people. Transhumanists believe that through fusing ourselves to brain interface memory chips and other human-enhancement technology, we will soon have implants to expand our senses and improve our cognitive functions. In other words, over the next decade or two, the fusion of man and machine may become a reality.The ultimate goal of science and technology is to produce individuals with extraordinarily high levels of intelligence, superhuman power, speed, and endurance, as well as noticeably longer lifespans. A strange undertaking given that population expansion is the main global source of environmental deterioration, greenhouse gas emissions, and consequently climate change. Mary Wollstonecraft Shelley (1797 – 1851) was well-aware of the latest inventions and scientific experiments. In 1816 she received an exclusive impetus for the implementation of her best ambitions. It was a famous summer when Mary Shelley, her lover Percy Bysshe Shelley, Lord George Gordon Byron, John William Polidori, and Claire Clairmont gathered at the Villa of Diodati near Geneva, Switzerland. Nasty weather made them all spend long days indoors, and Byron offered an idea to contest in writing horror stories. The friends were under the strong impact of German ghost stories, so the inspiration seemed to be easy to catch. However, the ghosts were not the subject of Shelley's interest. Instead, she insistently thought about a corpse reanimated by means of galvanism. *Frankenstein* explores a number of transhumanist ideas including the redefinition of the boundary between life and death and the revival of the dead; the production of beings that are hybrids of living and nonliving matter; and more. The monster's creation explores the production of human/sentient beings by mechanical construction, a scientific and engineered method of procreation. In a way, both Victor and his monster cross the boundaries of being human,but in deeply different ways. As a rational humanist and son of his time, Victor trusts in the possibilities of scientific progress, enforced by the education he received. His success is testified precisely by the contribution that his technological achievements represent for the progress of the human species by way of overcoming its limits and created a new species. On the other hand, Victor's offspring poses the ground for using those same technologies to transcend the limits of humanism and to postulate a new condition: the transhuman. In this regard, it is significant that Frankenstein's experiment comes along with a new understanding of the body, in particular of the corpse. In his act of creation, Frankenstein does not seem disturbed by decomposing bodies. Indeed, he explains: "Darkness had no effect upon my fancy, and a churchyard was to me merely the receptacle of bodies deprived of life, which, from being the seat of beauty and strength, had become food for the worm" (Shelley, 52). The idea of editing genes to alter a human is displayed in Frankenstein when Victor designs his monster to be more physically advanced than regular humans. When Victor witnesses the monster running from a distance he states, "As I said this I suddenly beheld the figure of a man, at some distance, advancing towards me with superhuman speed. He bounded over the crevices in the ice, among which I had walked with caution; his stature, also, as he approached, seemed to exceed that of man" (Shelley,). By means of contrasting the humanist ideal, like a cyborg, the Creature's physical diversity, its artificial origins and its ambiguous gender force it into a posthuman state. It is a fluid entity that subverts categorical fixities. It is neither male nor female, neither animal nor human, neither machine nor man. Rather, it is human, technology and animal all at once. As a result, Frankenstein's abhorrence and the rejection of his creature may epitomise the loathly aversion to any overhaul of humanist assumptions.

Mary Shelley's Frankenstein is sometimes admitted as the first true science fiction story. As early as the beginning of the 19th century the author managed to combine both practical and ethical criticism of transhumanist ideas. In the face of incomplete knowledge, the central character of the novel Victor Frankenstein carries out dangerous and groundbreaking experiments in a scientific laboratory. Its transhumanity is achieved only from a biological perspective and its failure resides in its attempts to become human rather than fully superseding and overcoming this definition. Moreover, designated as a monstrous, abhorrent creature from the very moment of its awakening, Frankenstein's creature seems to depart from the dream of perfectibility and faith in technological progress which the posthuman predicament encloses.

References

1. Burdett, Michael S: *Transhumanism and Transcendence*. Washington, D.C.: Georgetown University Press, 2011.
2. Nayar, Promod K: *Posthumanism,* Polity Press, UK, 2014
3. Shelley, Mary: *Frankenstein,* Fingerprint Publishing, 2015
4. González,Carretero, Margarita: *The Posthuman that could have been Mary Shelley's Creature,*GIECO, 2016

5. Rastelli, Giulia: *Reimagining the Human an Analysis of the Posthuman Turn in Mary Shelley, Kazuo Ishiguro and Jeanette Winterson*, Università Ca'Foscari Venezia,

6. http://pages.erau.edu/~andrewsa/sci_fi_projects_spring_2019 /Project_1/Birchler_Jack/HU_338-1/birchler.html

7. Transhumanism in Mary Shelley's *Frankenstein*- Book review, 2018

22. Research Exploration and the Transcendence of Research Methods and Methodology: An Interdisciplinary Perspective.

Dr. Rani Sarode

Associate Professor, Department of English and Language, Sandip University, Nashik MH, **India**

Abstract: This research endeavor embarks on a profound exploration of the ever-evolving landscape of scientific inquiry, emphasizing the transcendence of traditional research methods and methodologies. In a world where knowledge is dynamic and multidisciplinary, this study seeks to bridge the gap between conventional research paradigms and the innovative approaches necessitated by contemporary challenges. The initial phase of this research encompasses a comprehensive survey of the existing research methodologies across diverse fields, identifying commonalities and variations. It underscores the critical importance of interdisciplinary collaboration in addressing complex global issues. The study goes beyond the boundaries of specific research domains, recognizing the interconnectedness of knowledge and the need for holistic exploration.

The transcendence of research methods and methodologies involves the integration of cutting-edge technologies, data analytics, and emerging paradigms. It embraces a pluralistic approach that adapts to the ever-changing research landscape, ensuring that methods remain relevant and effective in the face of evolving research questions.

Through this research, we aim to pave the way for a more adaptable, inclusive, and agile approach to scientific inquiry. It emphasizes the symbiotic relationship between research methods and the transcendence of boundaries, showcasing the potential for greater discoveries and innovative solutions in a rapidly evolving world. This work serves as a beacon for researchers and scholars seeking to navigate the intricate web of interdisciplinary research, fostering a collective effort to transcend the limitations of methods and methodologies for the greater good of knowledge and society.

Keywords; exploration, transcendence, integration, researchers

Introduction

Research Exploration and the Transcendence of Research Methods and Methodology: An Interdisciplinary Perspective" is a comprehensive study that delves into the dynamic world of research methodologies and their evolution across various disciplines. In an era marked by the rapid advancement of knowledge and technology, this interdisciplinary perspective seeks to explore the ways in which research approaches have transcended traditional boundaries, offering a fresh outlook on how diverse fields converge and synergize in their pursuit of knowledge. This introduction sets the stage for an in-depth journey into the intricate tapestry of research methods, illuminating the transformative power of interdisciplinary collaboration in shaping the future of academia and innovation. Research exploration is a fundamental phase in the research process where a researcher investigates a specific topic or question to gather preliminary information and gain a better understanding of the subject. During this phase, researchers typically engage in activities such as literature reviews, data collection, and initial analysis. The primary goals of research exploration include identifying gaps in existing knowledge, formulating research questions, and selecting appropriate research methods and approaches. Research exploration is essential because it helps researchers establish a foundation for their study, ensuring that their research is well-informed and aligned with the current state of knowledge in the field. It allows researchers to refine their research objectives, define the scope of their study, and make informed decisions about the methodologies and tools they will use in the subsequent phases of the research process. Transcendence is a concept that signifies surpassing or going beyond the ordinary or usual limits. It is often used in various contexts, including philosophy, spirituality, and art. Here are a few common interpretations of transcendence: 1. Spiritual and Philosophical Context: In spirituality and philosophy, transcendence refers to the idea of surpassing the material world or the limitations of human existence. It can imply a state of existence or consciousness that goes beyond the physical realm, often associated with the divine or the metaphysical. 2. Art and Creativity: In the realm of art and creativity, transcendence may refer to the ability of a work of art, music, or literature to go beyond mere representation and evoke profound emotions or a sense of the sublime. It's about art's ability to transport the audience to a higher or deeper emotional and intellectual state.

3. Intellectual and Academic Context: In academic and intellectual discussions, transcendence can relate to the idea of going beyond established paradigms or boundaries. For example, "transcending traditional research methods" means moving beyond the limitations of standard research approaches to achieve new insights or perspectives. Overall, transcendence implies a sense of moving to a higher or more profound level, often involving a shift from the ordinary to the extraordinary, whether in a spiritual, artistic, or intellectual context. The transcendence of research methods refers to the idea of going beyond the conventional or traditional approaches to research. It involves seeking innovative and unconventional ways to conduct research that can yield fresh insights or address complex questions that may be difficult to tackle using standard methods.

Transcending research methods can manifest in several ways:

1. Interdisciplinary Approach: Researchers may transcend research methods by combining techniques and principles from different fields or disciplines. This interdisciplinary perspective allows for a more comprehensive and holistic understanding of a research topic.

2. Incorporating Advanced Technologies: With the rapid advancement of technology, researchers can transcend traditional methods by integrating cutting-edge tools and technologies, such as artificial

intelligence, big data analytics, and machine learning, to collect and analyzeata in ways that were previously unattainable.

3. Mixed Methods Research: Transcending research methods can also involve employing mixed methods, which combine qualitative and quantitative research approaches. This allows for a more comprehensive exploration of a research question by leveraging the strengths of both methodologies.

4. Innovative Data Sources: Researchers can transcend traditional data sources by exploring unconventional data streams, such as social media, sensor data, or real-time data sources, to gain new perspectives and insights.

5. Action Research and Participatory Methods: Transcending research methods can involve engaging with participants and stakeholders in a more collaborative and participatory manner, such as through action research, to address complex societal issues.

The transcendence of research methods is driven by the recognition that the world is evolving, and research methodologies need to evolve with it. By pushing the boundaries of traditional research methods, researchers can address contemporary challenges, discover novel solutions, and contribute to the advancement of knowledge in their respective fields. The transcendence of research methodologies signifies the evolution and expansion of the approaches, techniques, and paradigms used in the research process. It involves moving beyond the confines of traditional or established research methods to adapt, combine, or create new methodologies in response to the changing nature of research questions and the complex demands of contemporary research. Here are some key aspects of the transcendence of research methodologies: **Interdisciplinary Integration:** Researchers are increasingly transcending research methodologies by integrating insights and techniques from multiple disciplines. This interdisciplinary approach allows for a more comprehensive and multifaceted understanding of complex issues. **Mixed Methods Research:** Transcendence often involves adopting mixed methods research, where both quantitative and qualitative methodologies are used in tandem. This provides a more complete and nuanced perspective on research questions. Incorporating Advanced Technology: Advances in technology have opened up new avenues for research. Researchers can transcend traditional methodologies by leveraging advanced tools like artificial intelligence, machine learning, and data analytics for data collection, analysis, and visualization. **Action Research and Participatory Approaches:** Transcendence may involve embracing action research, participatory research, or community-based research methods, where researchers collaborate closely with participants or stakeholders to address real-world issues and co-create solutions.

Innovative Data Sources: Researchers are transcending research methodologies by exploring unconventional data sources, including social media data, sensor data, or open-source repositories, to gain fresh insights and perspectives. Longitudinal and Cross-Cultural Research: The transcendence of research methodologies often extends to conducting longitudinal studies or cross-cultural research, allowing for the examination of changes and comparisons over time and across diverse populations. **Qualitative Systematic Reviews:** In qualitative research, systematic reviews are being employed more frequently to transcend traditional literature reviews, offering a more rigorous and structured approach to synthesizing qualitative data. Transcending research methodologies is driven by the need to adapt to the complexities of the research landscape, address emerging questions, and enhance the validity and reliability of research findings. It encourages researchers to be innovative, flexible, and responsive to the ever-evolving demands of the academic and scientific community. Transcendence and research exploration within an interdisciplinary perspective are two intertwined concepts that reflect the evolving nature of research practices in response to the increasing complexity of contemporary issues and the dynamic intersection of various fields of study. **Transcendence in Research Methodologies:** Transcendence in research methodologies refers to the idea of moving beyond the confines of traditional research methods and paradigms. In an interdisciplinary context, this means breaking down the boundaries of discipline-specific research approaches and adopting innovative methods that draw from multiple fields. Researchers may transcend established methodologies by integrating insights, tools, and techniques from various disciplines to gain a more holistic understanding of complex problems. This approach enables a deeper exploration of research questions by combining the strengths of different methods. Research Exploration: Research exploration is the initial phase in the research process where researchers seek to understand a particular topic or question. It involves reviewing existing literature, gathering preliminary data, and formulating research objectives. In an interdisciplinary perspective, research exploration becomes a crucial step in identifying connections and overlaps between disciplines. It helps researchers recognize where traditional research methodologies may fall short in addressing multifaceted issues, prompting them to transcend these methods in pursuit of more comprehensive and innovative solutions. Interdisciplinary Collaboration: Transcendence and research exploration often go hand in hand within an interdisciplinary framework. Researchers from diverse fields collaborate to transcend the limitations of their individual methodologies. This collaborative approach enriches the research exploration process by promoting a holistic understanding of complex phenomena that transcend the boundaries of any single discipline. 4. Holistic Problem Solving: Interdisciplinary transcendence and research exploration allow for more holistic problem solving. By incorporating diverse perspectives and methodologies, researchers are better equipped to tackle multifaceted issues, providing comprehensive and innovative solutions that address the intricacies of the real world. In summary, the concept of transcendence in research methodologies and research exploration within an interdisciplinary perspective emphasizes the dynamic and adaptive nature of modern research. Researchers transcend traditional boundaries to explore complex topics, leveraging interdisciplinary collaboration and innovative methods to gain deeper insights and address the interconnected challenges of our time. This approach acknowledges that many real-world issues cannot be adequately addressed within the confines of a single discipline and calls for a more integrated and collaborative approach to research.

Conclusion

In conclusion, "Research Exploration and the Transcendence of Research Methods and Methodology: An Interdisciplinary Perspective" underscores the evolving nature of research practices in the contemporary academic landscape. This interdisciplinary exploration into the transcendence of research methods and methodologies has revealed the following key takeaways:

1. Interdisciplinary Collaboration: The study highlights the pivotal role of interdisciplinary collaboration in transcending conventional research methods. By bringing together diverse perspectives and methodologies, researchers can address complex, multifaceted questions more effectively.

2. Innovative Methodologies: The research has showcased the importance of embracing innovative research methodologies, often incorporating advanced technologies and mixed methods approaches. This adaptability allows researchers to capture a more comprehensive view of the issues at hand.

3. Holistic Problem Solving: The interdisciplinary perspective encourages holistic problem-solving. Researchers are better equipped to tackle real-world challenges by transcending the boundaries of

individual disciplines, recognizing the interconnectedness of contemporary issues.

4. Transcending Boundaries: The concept of transcendence extends beyond the confines of traditional research methodologies, emphasizing the need to adapt and innovate to address the complexities of modern research questions effectively.

In an era marked by rapid change and increasing interdisciplinary synergy, this research sheds light on the transformative power of transcending research methods. It reinforces the idea that a multidisciplinary approach not only enriches the research process but also holds the potential to drive innovative solutions and deeper insights into the multifaceted challenges of our time. As the academic and scientific community continues to evolve, embracing the transcendence of research methodologies remains a crucial endeavor for advancing knowledge and understanding in an increasingly interconnected world.

References

1. 1Sridhar, M. S. (2010). Introduction to Research Methodology: Problem Selection, Formulation and Research Design. United States: M. S. Sridhar, Lulu.

2. Kumar, M. (2020). C.R. Kothari: Research Methodology Methods and Techniques. India: New Age International (P) Limited. Academia.edu. Retrieved from https://www.academia.edu/33779875/C_R_Kothari_Research_Metho dology_Methods_and_Techniques.

3. Creswell, J. W. (2003). Research design: Qualitative, quantitative, and mixed Method approaches. A framework for design. Sage Publications, Inc. Second Edition.

4. Mishra, S.B., Alok, S. (2017). Handbook of Research Methodology. Educreation, 1.

5. Chawla, D. & Sodhi, N. (2011) "Research Methodology: Concepts and Cases" Vikas Publishing House PVT Ltd.

6. Vahidov, R. (2012). Design-type Research in Information Systems: Findings and Practices. United States: Information Science Reference.

7. Moissenko, F., Braicu C., Tomuleasa C., Berindan-Neagoe I. (2016) Types of Research Designs. In: Stefan D. (Eds) Cancer Research and Clinical Trials in Developing Countries. Springer, Cham. https://doi.org/10.1007/978-3-319-18443-2_3

8. Bickman, L., Rog, D.L. (2009). Applied Research Design: A Practical Approach. Handbook of Applied Social Research. SAGE Publications, London; 2009. 3- 46. 10.4135/9781483348858.n1